Agricultural Economics a An Introductic

Paul Brassley

Senior Lecturer in Agricultural Policy and History
Seale-Hayne Faculty of Agriculture, Food and Land Use
University of Plymouth

b
Blackwell
Science

To Angela, Charles and Caroline

© 1997 by Paul Brassley
Blackwell Science Ltd
Editorial Offices:
Osney Mead, Oxford OX2 0EL
25 John Street, London WC1N 2BL
23 Ainslie Place, Edinburgh EH3 6AJ
350 Main Street, Malden
  MA 02148 5018, USA
54 University Street, Carlton
  Victoria 3053, Australia

Other Editorial Offices:
Arnette Blackwell SA
  224, Boulevard Saint Germain
  75007 Paris, France

Blackwell Wissenschafts-Verlag GmbH
  Kurfürstendamm 57
  10707 Berlin, Germany

  Zehetnergasse 6
  A-1140 Wien
  Austria

First published 1997

Set in 10/13 pt Times
by DP Photosetting, Aylesbury, Bucks
Printed and bound in Great Britain by
Hartnolls Ltd, Bodmin, Cornwall

The Blackwell Science logo is a trade mark of
Blackwell Science Ltd, registered at the United
Kingdom Trade Marks Registry

DISTRIBUTORS

Marston Book Services Ltd
PO Box 269
Abingdon
Oxon OX14 4YN
(Orders: Tel: 01235 465500
         Fax: 01235 465555)

USA
Blackwell Science, Inc.
Commerce Place
350 Main Street
Malden, MA 02148 5018
(Orders: Tel: 800 759 6102
              617 388 8250
         Fax: 617 388 8255)

Canada
Copp Clark Professional
200 Adelaide Street, West, 3rd Floor
Toronto, Ontario M5H 1W7
(Orders: Tel: 416 597-1616
              800 815-9417
         Fax: 416 597-1617)

Australia
Blackwell Science Pty Ltd
54 University Street
Carlton, Victoria 3053
(Orders: Tel: 03 9347 0300
         Fax: 03 9347 5001)

A catalogue record for this title is available
from the British Library

ISBN 0-632-04137-4

Library of Congress
Cataloging-in-Publication Data
Brassley, Paul, 1946-
  Agricultural economics and the CAP: an
  introduction/Paul Brassley.
    p.   cm.
  Includes bibliographical references and
  index.
  ISBN 0-632-04137-4 (alk. paper)
  1. Agriculture and state – European
  Union countries.
  2. Agriculture – Economic aspects –
  European Union countries.
  I. Title.
  HD1918.B73   1996
  338.1′84–DC20
                                  96-38522
                                       CIP

CONTENTS

100102270X

PREFACE

For an industry which accounts for a small and decreasing proportion of the output of the European economy, agriculture gets a large slice of the European Union's budget and accounts for many of the political arguments which beset that organisation. Every family in the EU has its food prices determined by the Common Agricultural Policy (CAP), and the incomes of millions of farmers across Europe depend on the decisions made by those who control it. Yet few of them understand why the CAP exists or on what basic principles it is managed.

I have written this book for those who need to know, or would like to know, more about the CAP, and about the economics of the agricultural industry. Some good books have been written on these subjects (many of them are listed in Chapter 11), but most of them assume some knowledge of basic economic theory. This book does not. It is an elementary introduction, and so assumes *no* previous knowledge of economics on the part of the reader. The theory is inserted as necessary, and is as simple as possible. It is certainly not the last word on agricultural economics; I hope it may be a stimulating first word.

Although the book was originally written for students on an HND course in agriculture, it will also be useful as an introductory text for degree students of agriculture, agricultural economics, countryside management and environmental studies, and as a reference work for diploma and certificate courses in agriculture and related subjects, and for A level geographers.

One accumulates many debts in writing a book like this one. The greatest, perhaps, is to Seale-Hayne students over the past 20 years who have been exposed to most of the arguments herein, and have been patient with me as I groped my way towards what I wanted to say. My initial interest in the subject was stimulated by the teaching of John Ashton, Bill Cowie, Graham Ross and Martin Whitby, and by the necessity of being well prepared in arguments with David Harvey, Dai Pritchard, John Cotton, Martyn Ibbotson and Dick Blossom. I cannot help wondering what Brian Camm and John Medland would have made of this book but, sadly, I shall never know. Martyn Warren, Anita Jellings, John Kirk, John Usher, Martin Walbank, Geoff Hearnden, Ralph Hare and Mrs V. Walling all read the book in draft form and encouraged me to persist with it. Derek Shepherd was kind enough to read it all carefully, make many useful suggestions, and save me from some potentially very embarrassing howlers. Sarah Nightingale was constantly generous in sharing her up-to-the-minute knowledge with me. Shahzia Chaudhri at Blackwell Science was meticulous, imaginative and all that a good copy editor should be. I alone am responsible for the errors which remain. The book is dedicated to Angela, Charles and Caroline, who made it all possible and worthwhile.

Paul Brassley

Chapter 1

Introduction

This book attempts to explain why the European Union has a Common Agricultural Policy (CAP), and to outline the way in which that Policy works.

Agriculture is not unique in being an industry which attracts government support – in most countries the defence industry is similarly favoured and, in the EU, there is a Common Fisheries Policy to set alongside the CAP – but in most industrialised countries the prices of goods and the incomes of producers in most industries are not subject to specific government intervention. So what is it about agriculture that makes it unusual?

One obvious answer is that agriculture is a big and important industry. We can see just how big it is by measuring its contribution to the total output of the national economy (which economists call the *Gross Domestic Product* or GDP) or its share of the total national labour force (see Table 1.1).

Table 1.1 Agriculture in the UK economy, 1960–1995.

Year	Agriculture's contribution to GDP (£ million)	Agriculture's contribution to total GDP (%)	Number of people engaged in agriculture ('000)	Proportion of total workforce in agriculture (%)
1960	913*	4.0	1037	4.0**
1970	1126	2.6	750	3.0
1980	4114	2.1	650	2.6
1990	6436	1.3	565	2.1
1995	9004	1.5	533	2.1

* This is the figure for agriculture, forestry and fisheries all together.
** Estimate.

Sources: A. Burrell, B. Hill and J. Medland (1990) *Agrifacts*, pp. 4, 40 and 41, Harvester Wheatsheaf, Hemel Hempstead; H.F. Marks and D.K. Britton (1989) A *Hundred Years of British Food and Farming: A Statistical Survey*, p. 120, Taylor and Francis, London; MAFF (1996) *Agriculture in the UK 1995*, Table 1.1, HMSO, London.
It should be noted that there are detailed differences in the definition of the national workforce which are ignored here.

Table 1.1 shows that:

- The agricultural industry's share of the total national output in the UK is decreasing over time, which implies that as the economy grows other parts of it grow faster than agriculture. Thus, it is not surprising to find that agriculture uses a declining share of the labour force. In 1851, when the agricultural labour force was at its peak, 1.8 million people, representing about 23% of the total workforce, worked in agriculture. At the beginning of the twentieth century, the agricultural labour force was 1.3 million, 9% of the total workforce[1].
- Agriculture's share of the labour force is, and usually has been, higher than its share of the GDP, which implies that transferring a worker from agriculture to another sector of the economy would increase total GDP.

Table 1.2 shows that this pattern is repeated in other member states of the EU, but it also shows the variation from one member state to another. As late as the 1980s, the proportion of the labour force

Table 1.2   Agriculture in the EU economy, 1981–92.

	Agriculture's share of the labour force (%)		Agriculture's share of Gross Value Added (%)	
	1981	1992	1981	1992
EU-12	9	6	9	3
Belgium	3	3	2	2
Denmark	8	5	7	3
Germany	5	4	2	2
Greece	31	22	17	15
Spain	19	10	6	4
France	8	6	5	3
Ireland	17	14	13	8
Italy	13	8	6	4
Luxembourg	5	3	0	2
Netherlands	5	4	4	4
Portugal	27	11	10	7
UK	3	2	2	2
Austria	10	7	—	3
Finland	13	9	—	6
Sweden	6	3	—	3

Source: Eurostat (1995) *Agriculture: Statistical Yearbook, 1995*, pp. 30, 31, Brussels.
EU-12 is the average for the 12 European Union nations.

employed in agriculture in Greece and Portugal was similar to levels which have not been seen in the UK since the middle of the nineteenth century. On the other hand, agriculture's contribution to the Gross Value Added was similar to UK levels in Luxembourg, Belgium and Germany.

From Tables 1.1 and 1.2 it is apparent that although agriculture is a big industry in the EU, it is also declining in importance, if importance is measured simply by its contribution to the national economies of the member states (of course, many people would argue that this omits such important considerations as agriculture's effect on the landscape and environment, and on rural society as a whole). So does that mean that farming gets a common EU policy just because it is a declining industry?

Much of the rest of this book is concerned with answering this question, but it may help at this point to outline the argument so that the relationship between the individual chapters is made clearer. This is what Fig. 1.1 attempts to do.

How to use this book

Each of the following chapters, except for Chapter 2, is concerned with one of the boxes making up Fig. 1.1. Chapter 2 deals with the economist's theory of demand, supply and the price mechanism, a bit of basic theory which you will find very useful in understanding the rest of the book. It's the only theoretical chapter because other aspects of economic theory are introduced only at the point at which they are needed to explain the arguments. If you already know something about the economics of the market you can probably skip Chapter 2; if you don't, try to work your way through it. You will be able to understand much of the rest of the book without it, but you will find that it all makes more sense if you know something about how markets work. After that, you can see how the chapters relate to the argument outlined in Fig. 1.1:

- Chapter 3 explains why the demand for agricultural products is static.
- Chapter 4 examines long-term changes in production in the UK and the EU.
- Chapter 5 explains the factors that affect the supply of agricultural products.
- Chapter 6 is concerned with the inputs used in agriculture.
- Chapter 7 looks at the particular problems of trade in agricultural products.

The *demand* for agricultural products is fairly static in most industrialised countries because most people have enough to eat, and so, as they get richer, their additional expenditure goes on things other than food, from additional health care in the cradle to more elaborate funerals at the grave. Moreover, there is little extra demand to be had from more mouths to feed, because the population in most of these countries grows only slowly. The food processing industry is constantly developing new products, but few of them require many extra raw materials, and that is what agriculture mostly produces – raw materials for the food processing and retailing industry.

The *supply* of agricultural products in industrialised countries tends to increase over time, largely because farmers introduce new technology, which tends to be output-increasing.

Some countries produce this extra supply cheaply enough for it to be sold on the world market without the aid of export subsidies, but for most member states of the EU and most of the products they produce, this is not the case.

Supply also fluctuates from one year to another in response to variations in the weather and the incidence of pests and diseases.

So, faced with static demand and supplies which increase over the long term and fluctuate over the short term, the prices of agricultural products tend to fluctuate in the short term and decrease (after allowing for inflation) over the long term, in a market in which there is no governmental intervention. Consequently, the income of the agricultural industry as a whole tends to decrease....

.... which might not matter if there were fewer farmers with less land to share the decreasing income but, in some countries, many of the farmers remain in agriculture unless very hard pressed, and so does most of the land.

So, in the absence of government support, farm incomes tend to fluctuate in the short term and decrease in real terms in the long term....

....which means that governments have to decide whether or not to support farm incomes. There are some good reasons for doing so, and some good reasons for not doing so, and over the past century industrialised countries have sometimes decided in favour of support and sometimes against it, but since World War II most have had some form of support.

In the EU, that means the Common Agricultural Policy (CAP).

Fig. 1.1   An outline of the agricultural problem.

- Chapter 8 examines the arguments for and against farm income support.
- Chapter 9 sets the support arguments in their historical and political context.
- Chapter 10 is what it's all been leading to: the CAP and how it works.
- Chapter 11 reminds you that this is only an introduction to the subject and suggests some further reading if you want to know more about it.

You will find that each chapter contains questions for you to answer as you work your way through. The purpose of most of these is to enable you to check that you have understood what you have read, but there are just a few which are designed to provoke you into thinking further about the issues raised.

If you are using this book as part of a taught course, your teacher or lecturer will tell you which bits to read, and when. If you are working on your own, you might like to bear these points in mind:

- Don't feel that you have to begin at page 1 and read through to the end. Flick through the book to get a rough idea of the contents, and when you read each chapter, skip through it quickly to get a general idea of what it's about before settling down to read it in detail.
- Keep referring back to Fig. 1.1 to see how any particular chapter relates to the overall argument.
- You might find it helpful to write *brief* notes on the main points of each section as you finish reading it. Putting things into your own words helps you to check that you have understood what you have read. The questions are there for the same purpose.

Note

1. For more information, see A. Burrell, B. Hill and J. Medland (1990) *Agrifacts*, p. 41, Harvester Wheatsheaf, Hemel Hempstead.

Chapter 2
How Markets Work

2.1 Introduction

'1998 should be an excellent season for potato growers, with reduced acreages and total production inducing higher prices', a report on agricultural markets might say. For example, in one year, the potato harvest might produce 7 million tonnes from about 148 000 hectares, and in December of that year potatoes sell for about £70 per tonne. In the following year, plantings decline to about 142 000 hectares, producing 6.3 million tonnes, and prices rise to nearly £160 per tonne by the end of the year.

What happened to the potato market in the example, and how can we explain the relationship between what farmers decide to grow and the price they receive for their products?

In this chapter, we are concerned with some of the basic ideas of markets, how they work, and how they determine prices.

2.2 How demand and supply affect price

Everybody knows that price is determined by demand and supply. If buyers want more of something, and no more is available, the people who will get more are those who are willing and able to pay more. If producers supply more, and no more is wanted, the price will fall. This is the sort of thing you can see happening in the market for fat lambs. When buyers have plenty of orders for exports to the continent, prices will rise; if the export market is disrupted, by anything from bad weather in the Channel to farmers' riots in Normandy or animal rights activists at the ports, prices will fall.

These ideas can be visualised by using Fig. 2.1, in which one line (labelled D) represents demand and the other (labelled S) represents supply. These two lines are called the *demand and supply curves* (even

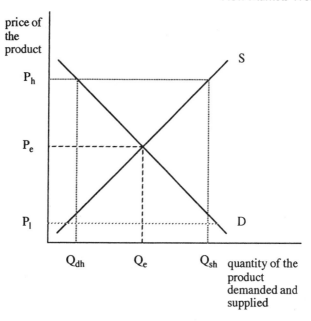

Fig. 2.1   The demand and supply curves.

though they may be drawn as straight lines – it's just one of the little foibles of economists!).

The demand and supply curve in Fig. 2.1 shows the relationship between the price of a product and the quantity demanded. The supply curve shows the relationship between the price of a product and the quantity supplied. If the price is set at P_h (high price), suppliers will want to supply more to the market (Q_{sh}) than consumers wish to buy (Q_{dh}). If the price is reduced to P_l (lower price), demand will exceed supply. At such low prices, consumers who want to buy more of the product may seek out suppliers and offer to pay higher prices. Clearly, there is only one price at which the quantity that consumers wish to buy is the same as the quantity that suppliers want to sell, and that is P_e, the *equilibrium price* (with its corresponding equilibrium quantity demanded and supplied, Q_e). This is the point where the curves cross and, because the quantities demanded and supplied are the same, there is no reason for the price to rise or fall.

Now try going over the same idea again, but this time using an example from the real world.

The wheat market

Supply

If the price is high, wheat production is profitable and lots of farmers will want to produce wheat. They may use more fertilisers and sow less barley in order to make land available for wheat. Consequently, the quantity of wheat supplied to the market will increase.

If the price is low, farmers will find it more profitable to grow barley, oats or some other alternative crop, so the quantity of wheat supplied to the market will decrease

Demand

If the price is high, feed compounders, for example, will look for alternatives to wheat for use in livestock rations. They may buy barley or import manioc to incorporate into their concentrates. Consequently, the quantity of wheat demanded on the wheat market will decrease.

If the price is low, feed compounders will use wheat in livestock rations in preference to alternatives, which are now comparatively more expensive then wheat, and so the quantity of wheat demanded by the market will increase.

➪ **Questions**_____

The demand/supply relationship is the model on which our understanding of markets is based, so it is vital that you understand it. To make sure that you do, try drawing the demand and supply curves based on the data below, and then answer the following:

Milk price (p per litre)	Quantity demanded (million litres per week)	Quantity supplied (million litres per week)
15	320	180
16	310	206
17	302	232
18	293	260
19	285	285
20	274	312
21	266	338
22	256	364

1   What would be the size of the shortage or surplus at a milk price of 15p per litre (ppl)?

2 What would be the size of the shortage or surplus at a milk price of 22p per litre?

3 What is the equilibrium price and quantity?

4 If the demand for milk increased by 40 million litres per week at all prices, what would be the new equilibrium price and quantity? (You will probably find it helpful to plot the new figures on your graph.)

2.3 What affects the demand for a product?

All sorts of things affect the demand for a product, but economists usually reckon that the following are the most important:

☐ *The price of the product:* The higher the price of the product, the smaller the quantity which will be demanded.

○ *The price of substitutes for the product:* Margarine is a substitute for butter, and if the price of margarine goes up, we would expect margarine buyers to think about buying butter instead. Conversely, if the price of margarine goes down, butter buyers may buy margarine instead.

○ *The price of complementary products:* Some products just naturally go together, like beer and crisps, lamb and mint sauce, twine and balers, videotapes and video recorders. Thus, if more video recorders are bought because they become cheaper, it's also likely that more videotapes will be bought to play in them. But the relationship may not work equally well both ways. We may buy lots of mint sauce because lamb is cheap and we are eating plenty of it, but a sudden decrease in the price of mint sauce is unlikely to send us out shopping for lamb.

○ *The incomes of consumers:* The richer you become the more you are likely to buy. However, it is important to remember that this does not work equally well for all products. The very rich may buy large quantities of smoked salmon and fine wines, but they may not necessarily buy many more potatoes than poorer people.

○ *The tastes of consumers:* These vary from one consumer to another, and each individual's tastes may change over time. The change in tastes away from cooked breakfasts, and government campaigns to make people aware of the effects of saturated fats in the diet have had a big effect on the demand for bacon, butter and red meat.

○ *Expectations of future price rises* may encourage people to go out and buy when prices are beginning to rise, which is not what you would expect if their behaviour was just controlled by current price.

O   *The number of consumers*, or, in other words, the population. The more people there are, the greater the size of the total market and the greater demand will be, although it is important to identify people who really are potential buyers. Thus, there are millions of food consumers, but only thousands of potential buyers of pedigree livestock.

(You will notice that one of the factors in this list is marked with a □, and the rest by circles (○). The reason for this is explained in section 2.5 on how the curves work.)

2.4   What affects the supply of a product?

In Chapter 5 we shall look in more detail at the things that affect supply, but for the moment we can produce a list of the following factors:

□   *The price of the product:* As prices rise, more and more producers will believe that they can make money by producing the product, and existing producers will also find it worthwhile to produce more. Thus, when cereal prices are high, farmers on poorer, higher land may be tempted to grow barley, and as prices fall that land will go back into grass.

O   *Costs of production:* As costs rise, profits fall for any given price level so producers attempt to cut back on inputs, which may have the effect of reducing output. So if fertiliser became very expensive, farmers might decide to use less of it and so yields would fall.

O   *The state of technology:* Technical changes, such as the development of new crop varieties or pesticides, can increase the yield resulting from a given level of inputs.

O   *The profitability of other products which the firm produces:* If the price of barley rises relative to the price of wheat, farmers will probably decide to grow a bit more barley and a bit less wheat, and vice versa.

O   *The profitability of products produced together:* Some things are by-products, so when the producers decide more of one they inevitably produce more of the other, and vice versa. Thus, wool production rises as more ewes are kept to produce fat lambs, and when dairy cow numbers fell after the imposition of milk quotas there were fewer calves available for fattening.

O   *The goals, or objectives, of the producers:* These are generally assumed to be to make as much profit as possible, but may not always be so. For example, some part-time farmers may adopt a system which gives them time to do another job away from the farm.

○ *Expectations of future price changes* may encourage producers to increase or decrease output before the market price actually moves.

○ *The unpredictable:* Weather, disease, war, strikes, earthquakes, fires, floods, and the tendency of any complex piece of machinery to go wrong shortly before you really need it or shortly after you get angry with it, will all affect the quantity supplied to the market in the short term.

⇨ **Questions**_____

Consider the market for fat pigs. In which of the categories above would the following changes appear? (For example, a prediction of a shortage of pigmeat would affect expectations of future price rises, which is one of the few factors affecting both demand and supply.)

5 A decline in the popularity of cooked breakfasts?
6 An increase in the price of chicken?
7 An increase in the price of fishmeal?

2.5 How demand and supply curves work

Two things can happen, depending on the factor that changes:

(1) There can be *movement along* the curve if the price changes, i.e. if there is a change in one of the factors marked with a □ in the list of factors affecting demand and supply.

For example, if prices rise, from P_1 to P_2, we move up the curves. The quantity demanded (Q_d) and the quantity supplied (Q_s) increase so we go from Q_{d1} to Q_{d2}, and from Q_{s1} to Q_{s2} (see Fig. 2.2).

(2) The curve can *shift*, either to the right or to the left if anything other than the price changes, i.e. if there is a change in one of the factors marked with a ○ in the list of factors affecting demand and supply.

For example, if incomes rise, the demand curve will shift to the right, from D_1 to D_2. This is because, as people get richer, they normally buy more and so, at any given price, P, the quantity, Q, demanded is greater. This is illustrated in Fig. 2.3.

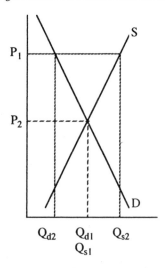

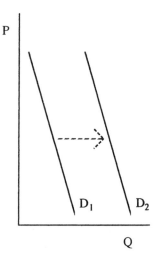

Fig. 2.2 The effect of a price increase   **Fig. 2.3** The effect of an income
on quantities demanded and supplied.   increase on the demand curve.

⇨ **Questions**_____

Now draw the diagrams again, for different circumstances:

8 Draw Fig. 2.2 again, this time showing what happens if the price falls.
9 Draw Fig. 2.3 again, this time showing what happens if there is a
decrease in income.
10 Draw Fig. 2.2 again, showing what happens when some new output-
increasing technology is introduced.

2.6 Combining demand and supply again

Now we can go back to the idea of the price being determined by
demand and supply. We can go further, because we now know what
affects demand and supply and how we can illustrate the process using
the demand and supply curves. There is nothing magic about them.
They are just a way of helping us to visualise what is going on in the
market, a way of helping us to think clearly.

We know that the demand curve shifts to the right if there is an
increase in the incomes of consumers. If nothing else changes, what will
happen to the price of the product as a result? Fig. 2.4 illustrates the
process.

We begin with demand curve D_1, giving us a price P_1 and quantity Q_1
demanded and supplied. If we then shift the demand curve to the right,
to D_2, as a result of the income increase, the price will rise to P_2 and the

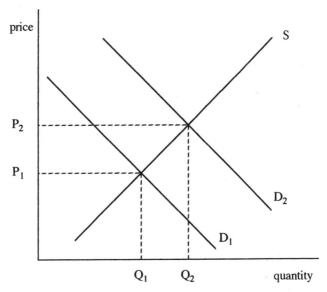

Fig. 2.4 The effect on price of an increase in consumers' income.

quantity demanded and supplied will rise to Q_2 as producers supply more to the market in response to the increased price.

If we then shift the *supply* curve to the right, to S_2, because, say, costs of production fall, we can see in Fig. 2.5 that the new price will be P_3 and the quantity on the market will be Q_3.

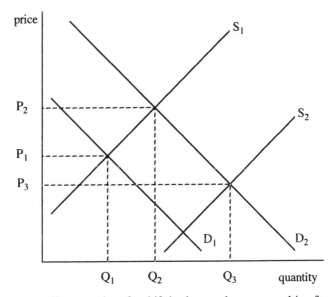

Fig. 2.5 The effect on price of a shift in the supply curve resulting from a decrease in costs of production.

You can see how we can use the demand and supply curves to work out what will happen to the price of a product and the quantity on the market when a change occurs to affect the market. Use the following checklist to help yourself to ask the right questions:

(1)   Does the change affect demand or supply, or both? (Only one factor affects both. If you're not sure, go back to the lists in sections 2.2 and 2.3 to check.)
(2)   Does the change bring about a *movement along* the curves (only one will do this)? If this is the case, is it going up or down?
(3)   Does the change cause a *shift* in the curve? If this is the case, is it a shift to the left or the right?

Draw the demand and supply curves according to how you answer the questions.

Now we've reached the point at which it might be useful to try all this theory on a few examples. Here's one to give you an idea of what goes on:

The asparagus market

What will happen to the price of asparagus following the landing of a spaceship full of little green extra-terrestrials who live on asparagus? If we go through the checklist we discover that:

(1)   This spaceship full of little green ETs will increase the *number of consumers* of asparagus, and if we look on the list of factors affecting demand and supply we find that this is on the demand list, with a circle (○) next to it.
(2)   So when we ask if it produces a *movement along* the demand curve, or a *shift*, we remember the rule that says that all factors marked with a circle produce a shift.
(3)   All we have to sort out then is whether the shift will be to the *left* or to the *right*. Since the number of consumers will have increased, more will be demanded at any given price, so the demand curve will shift to the right.

When we draw that on the diagram (Fig. 2.6), we see that the effect of increasing demand from D_1 to D_2 is to *increase* the price of asparagus and the quantity on the market (P_1 to P_2 and Q_1 to Q_2).

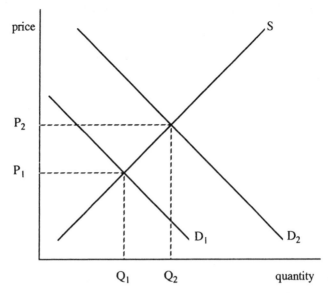

Fig. 2.6 The effect of an increase in population on the price of asparagus.

⇨ **Questions**_____

Consider the market for milk. What will happen to the milk price and the quantity on the market (assume that milk quotas have been abolished) if the following things occur?

11 Consumers' incomes increase?

12 The population increases?

13 The price of fat lamb increases?

14 The price of nitrogenous fertiliser increases?

15 Milk Marque raises the milk price?

16 Better breeding increases the milk yield of the average cow?

17 Margarine prices decrease?

18 The government produces a report saying that saturated fats produce health risks?

19 There is a drought in the summer?

20 The value of the £ changes so that the intervention price of butter decreases?

When you have done these questions, you should be able to work out the *direction* in which price changes – up or down. The next thing is to work out by *how much* the price will change. Are we dealing with a big change or a small one?

2.7 Big and small changes: the idea of *elasticity*

The price of potatoes probably has to rise a lot in Britain before people cut back significantly on their consumption, whereas if lamb prices rise consumers may switch to beef, pork or chicken for their weekend roast (if they still have a weekend roast – but that's another issue). This behaviour has some important implications. It means, for example, that if there is a shortage of potatoes, prices will rise a lot because consumers will tend to keep on buying them despite the initial price increases. Equally, if there is a glut of potatoes, consumers will not be tempted by low prices to buy lots more of them. The demand for potatoes is thus said to be *inelastic* with respect to price (see Fig. 2.7). A large drop in price from P_1 to P_2 has only a small effect on the quantity, Q, demanded. In contrast, a consumer who is equally fond of both roast beef and roast pork may switch from one to the other just because one happens to be a little bit cheaper. We can see the effect of this in Fig. 2.8, where a small decrease in the price of beef (from P_1 to P_2) might produce a big increase in the quantity demanded (Q_1 to Q_2), so that the demand for beef is said to be *elastic* with respect to price.

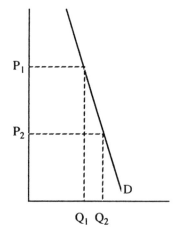
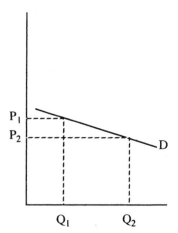

Fig. 2.7 An *inelastic* demand curve.   **Fig. 2.8** An *elastic* demand curve.

There are four types of elasticity with which you need to be familiar:

(1) *Price elasticity of demand,* which measures the extent to which a price change affects the quantity demanded.
(2) *Cross-price elasticity of demand,* which measures the extent to which a change in the price of one product affects the quantity demanded of another product (for example, if the price of mar-

garine goes down, will the quantity of butter demanded go down a little bit or a lot?).

(3) *Income elasticity of demand*, which measures the extent to which a change in consumers' incomes affects the demand for a product (if people get richer will they want lots more liquid milk or only a little bit more?).

(4) *Price elasticity of supply*, which measures the extent to which a change in the price of the product persuades producers to supply more or less to the market (if the price of wheat goes up, how much more will farmers produce?).

We can calculate a value for *price elasticity of demand* using the formula

$$\text{Price elasticity of demand} = \frac{\text{Percentage change in quantity demanded}}{\text{Percentage change in price}}$$

Remember five things about this formula:

(1) In all elasticity formulae, the change in *quantity* always goes on top.

(2) In all elasticity formulae the changes are always measured in *percentage* terms. If you wonder why, think about calculating price elasticities of demand for houses and newspapers. An extra 100 houses for sale in a small town would be a big proportion of the total on sale, whereas an extra 100 newspapers in daily sales of several million would pass unnoticed. Similarly with price: 10p extra on the price of a house would hardly affect the likelihood of selling it, but circulation wars have been fought over 10p changes in newspaper prices.

(3) The result of the price elasticity calculation will always be *negative*, because an increase in price will produce a decrease in quantity demanded, and vice versa. The rules of arithmetic demand that if you divide a positive quantity by a negative, or a negative by a positive, the result will be negative.

(4) So if the result of the elasticity calculation is between 0 and −1, it means that a 1% change in price will produce less than a 1% change in the quantity demanded, and the product is said to be *price inelastic*. If the price elasticity is −2 or −3, or some bigger negative number, the product is said to be *price elastic*, meaning that a 1% change in price will produce more than a 1% change in quantity demanded.

(5) Price inelastic products tend to be necessities for which there are no good substitutes (such as bread, potatoes and liquid milk), whereas the demand for luxuries (fillet steak perhaps?), or products which can easily be replaced by something else, will be more

responsive to price. Thus, for pig producers, the demand for concentrates as a whole will be price inelastic, but the demand for a particular brand will have a much higher price elasticity. This is because (despite anything the firms' salesmen may say) one firm's fattening ration is probably a pretty good substitute for another's, so one firm's price increase could have a significant effect on another's sales.

The formulae for the other types of elasticity are similar to the price elasticity formula:

$$\text{Cross-price elasticity of demand} = \frac{\text{Percentage change in quantity of product A (e.g. butter)}}{\text{Percentage change in price of product B (e.g. margarine)}}$$

The cross-price elasticity will be *positive* for substitutes like butter and margarine, because an increase in the margarine price will produce an increase in the quantity of butter demanded. If the two products are good substitutes for each other a small change in the price of one may have a big effect on the demand for the other (as in the pig feed example above), but it would take a pretty big change in tractor prices to increase the demand for carthorses.

$$\text{Income elasticity of demand} = \frac{\text{Percentage change in quantity demanded}}{\text{Percentage change in consumers' incomes}}$$

Again, the income elasticity will be *positive*, because an increase in income will normally produce an increase in quantity demanded (although there are some exceptions to this rule which are important for agriculture, as we shall see in the next section of the course). Necessities have low income elasticities (typically less than 1, so if incomes increase by 1%, spending on a necessity will usually rise by less than 1%) and luxuries have high income elasticities (a 1% income increase may produce a 2 or 3% rise in spending on tourism).

$$\text{Price elasticity of supply} = \frac{\text{Percentage increase in quantity } supplied}{\text{Percentage change in price}}$$

Again, the supply elasticity will be *positive* because an increase in price will encourage suppliers to produce more, and vice versa. It will also increase over time, because it usually takes time for producers to adjust to a price change. Additionally, the supply elasticity depends upon how much it costs to produce a bit more: if firms have spare capacity or can produce more without big increases in labour costs, they may respond rapidly to a price increase; if they can't get hold of raw materials, or are already working flat out, the supply elasticity may be low.

⇨ **Questions**_____

Now see if you can answer the following about elasticity:

21 Is the price elasticity of demand likely to be higher for steak or for meat products in general?

22 The cross-price elasticity for substitutes is positive, but is it likely to be positive or negative for complementary products such as beer and crisps?

23 Is the income elasticity of demand likely to be higher for manufactured foods such as frozen chocolate gateau or for unprocessed foods such as loose ware potatoes?

24 Would you expect the price elasticity of supply for timber to be higher or lower than that for wheat?

25 The price of feed wheat increases by 5% and, as a result, the quantity demanded by compound feed manufacturers decreases by 8% (these figures are imaginary). What is the price elasticity of demand for feed wheat?

26 If the price elasticity of demand for home-baked cakes sold in Little Drivel in the Mush village store and post office is –1.5, and prices are increased by 20%, what will be the percentage change in the quantity demanded?

27 When consumer incomes increase by 3%, what will be the percentage change in the demand for food raw materials if their income elasticity of demand is 0.2?

28 When the price of wheat increases by 5%, farmers produce 1.5% more wheat. What is the price elasticity of supply of wheat?

29 If a product has a price elasticity of demand between 0 and –1, and its price increases, will the total revenue (i.e. the price of the produce × the quantity sold) to the producer increase or decrease?

2.8 Markets in the real world

It is important to remember that the demand and supply curves are no more than ways of trying to explain what happens in the real world, and that they are sometimes simpler than reality. There are several problems which affect real markets which we haven't discussed so far.

One of the most important is *time*. Strictly speaking, demand and supply curves deal with one particular moment of time. We relax this restriction when we shift a demand curve in response to, say, a change in consumer income, but that's about as far as it goes. We implicitly assume, for example, that production is more or less instantaneous.

That may be quite reasonable for some products but, obviously, changes in taste, such as the decline in the popularity of the cooked breakfast, may take place over decades. Oaks which were planted in response to high timber prices at the time of Trafalgar (1805) would probably now make good shipbuilding timber, but for some reason the Royal Navy does not seem to be buying them.

There are also examples of *market failure*, when the market *doesn't* produce what society wants. The market will ensure that farmers are paid for the food that they produce but not for the pleasant landscapes that hundreds of years of farming have created. And neither will the market charge them for producing things that society doesn't want, such as silage effluent or pesticide residues. There are several reasons for market failure, and various ways of coping with them.

(1) *Uncertainty.* The demand and supply model assumes that a beef producer knows what the price of fat cattle will be when they are ready for market; in the real world the decision to produce them may be taken two years before and prices can change a lot in that time. In some cases, mechanisms have evolved which attempt to deal with the problem: forward contracts and future markets are obvious examples.

(2) *Externalities* arise because the actions of a firm may have effects which are external to the firm, but have some impact on the rest of society. They may either be costs or benefits:
  *   *External costs* are created when, for example, a farmer discharges a pollutant into a stream and kills the fish that the angler would otherwise derive pleasure from catching. The market mechanism, left to itself, will do nothing about this, so often society will create some legal constraint to prevent the pollution by fining the polluter.
  *   *External benefits* have resulted, as mentioned above, from the activities of farmers in creating the landscapes that tourists enjoy. Such landscapes are a by-product, and although the market mechanism ensures that farmers are paid for milk and wheat and so on, it pays nothing for the scenery. Consequently, the profit maximising farmer will concentrate on producing what pays, and forget about what doesn't, which may not be what society as a whole wants. One way of overcoming this is to pay farmers to farm in ways which maintain the preferred landscapes, and that is the purpose of schemes such as Environmentally Sensitive Areas.

(3) *Public goods* are those which are:
  *   *Non-excludable*, which means that you can't provide them for

one person without them being available for everybody: imagine the problems of erecting flood defences on the Lincolnshire coast which would only protect one farm.

- *Non-rival*, which means that their consumption by one person doesn't prevent their enjoyment by somebody else: if I get pleasure from looking at a landscape, it doesn't prevent you from enjoying the sight of the same landscape.

Therefore, the market mechanism will either supply less public goods than society would like, or not supply them at all. That is why they are normally supplied by national or local government. Examples range from the activities of Internal Drainage Boards to the provision of National Parks.

(4) *Market power* often results from a firm controlling a high proportion of the output of a market. Consequently, it is able to influence the quantity being bought or sold on the market, and so affect the price. The ultimate market power is that exercised by the monopoly seller or buyer. The old Milk Marketing Board was a monopoly buyer of milk, and so milk producers had to accept the Board's price, even though some large producers located near urban markets for liquid milk felt that they might have been able to get better prices if they had been able to operate on their own. Farmers often buy from and sell to firms which have a significant proportion of the total market (e.g. there are only a few fertiliser and feeding stuffs manufacturers, and a small number of supermarkets control most of grocery sales), so in the absence of government intervention they would have little influence over both input and output prices. It is the function of the Monopolies and Mergers Commission to decide whether or not effective competition exists in an industry.

2.9 Markets and other aspects of economics

We have now examined markets and the way in which they work in some detail. Is that all economics is about? Are the economics correspondents in the newspapers and on TV simply arguing about the way markets work? Clearly, they are not, although markets are central to any discussion of the economy. So what is economics all about? Here is a definition:

> Economics is about the way *choices* are made about the use of *scarce resources* in order to satisfy *objectives*.

Let us look in more detail at the meaning, in the context of economics, of the words in italics.

Choices: People have to make choices. You had to decide whether to buy this book, or steal it (in which case, thank you for the implied compliment, but wouldn't it have been more sincere if expressed through your chequebook?), or join a syndicate to buy it jointly, or borrow it from the library. Or perhaps you are reading this while browsing in a bookshop and in a few moments you will return this book to the shelf and buy something else. In each case, you will have made a choice, and economists are interested in the choices people make, and in why they make them.

Scarce: Most resources are scarce, in that they are not available in infinite quantities at no cost. There are plenty of people in the world, for example, but try getting a plumber on Sunday. One of the few resources which isn't normally scarce is air but it becomes so in a submarine, and clean, fresh air becomes scarce in city centres on hot sunny days.

Resources: are the things we use to make the things we want. Another word for them is inputs, which we shall discuss in the context of agriculture in Chapter 6. Inputs can be divided into land, labour, capital and management. We can use these four together to produce a product. For example, we can take some land, seed, a plough and a harrow pulled by a bullock, a sickle and some labour to produce wheat. Or we can use the same land, some seed, seed dressing, fertiliser, several sprays, a plough, a set of harrows, a combine drill, a sprayer, a combine harvester, a tractor and trailer and rather less labour, and also produce wheat. So resources can be substituted for each other up to a point.

Objectives: Individual people and firms, and even whole societies, all have objectives. Firms may wish to maximise their profits, or the rate at which they grow. Individuals may wish to be happy, rich, famous, or on holiday in France during the month of August. What are your objectives?

So now, perhaps, you can see why we have taken a chapter to examine in some detail the workings of the market. The market is a mechanism for enabling decisions to be made about the use of scarce resources in order to satisfy objectives. It is not the only available mechanism. We don't use it for deciding on the number of soldiers or police officers we want, and only some people use the market to acquire education and health services. In fact, insofar as farm incomes are supported by the European Union, it appears that the quantity of resources used by the agricultural industry is not entirely controlled by the market.

This study of the activities of producers and consumers in the market forms the subject matter of *microeconomics*. Economists also study the way in which national and international economies work, and examine problems such as employment, inflation, interest rates, and so on. These are matters of *macroeconomics*. With a few exceptions, most of the issues discussed in this book use microeconomic concepts.

2.10 Summary

- Price is determined by the interaction of demand and supply.
- Both demand and supply are affected by several different factors, and it is important to remember which affects demand and which supply. Only price affects them both.
- Changes in price produce movements along the demand and supply curves; changes in any of the other factors produce shifts in the curves.
- Elasticities measure the extent to which prices and quantities change.
- There are some real-world problems which the demand/supply model doesn't really cope with.
- The study of markets is a vital part of microeconomics.

Chapter 3
The Demand for Agricultural Products

3.1   Introduction: applying demand theory to agriculture

In Chapter 2 we looked at the way markets work, and saw how they were affected by changes in the demand for and supply of a product. This chapter applies what we discovered about demand to the market for agricultural products. In other words, it is concerned with the particular characteristics of agricultural products, and the way in which those characteristics affect demand.

One of the first questions that arises, therefore, is 'What does agriculture produce?' At first sight this is so straightforward as to be hardly worth asking. Obviously, agriculture produces food: potatoes, beef, bacon, lamb, wheat to make bread, barley to make beer, milk, apples, and so on. But think about this list again.

- How many of these products could you buy from an ordinary farm? The potatoes and the apples perhaps, but the others all require some processing before they become food for the consumer to buy, and very often the potatoes will be washed and pre-packed and the apples will be graded before most consumers buy them.
- How many of these products are not sold to consumers at all but consumed on the farm or sold to other farmers? Some varieties of wheat will go into bread or biscuits but much of the wheat produced in Europe is soft feed wheat which will be sold to animal feed compounders, and the same applies to barley.
- Remember that lambs come from ewes which will produce wool every year, so agriculture also produces some non-food products; vegetable fibre crops such as flax and hemp are other examples.
- Some farmers, especially those in the more beautiful but less agriculturally productive parts of Europe, cash in on people's desire to spend their holidays in the country and sell camping and caravanning facilities and holiday cottages. Thus, they are selling some of their products to the tourist industry.

Mostly, therefore, *agriculture produces raw materials for the food industry*, but it is important to remember the exceptions to this generalisation. Agricultural economists know quite a bit about the demand for food in the UK because, each year since 1940, the Ministry of Agriculture, Fisheries and Food (MAFF) has carried out a survey of food consumers' behaviour (published annually since 1992 as *The National Food Survey*, and before that as *Household Food Consumption and Expenditure*), but rather less about the behaviour of the firms that buy agricultural raw materials directly from farmers. However, since the demand for the raw materials is derived from the demand for the final product, it is important to find out about the way in which food consumers behave if we are to understand the influences acting on the market for farm products.

From the list of factors affecting the demand for a product which we examined in the previous chapter, we will select the following to examine more closely:

- The effect of price changes
- The effect of changes in the price of substitute products
- The effect of income changes
- The effect of changes in taste
- The effect of population changes.

3.2   Price changes and demand

Price changes produce movements along the demand curve. As potatoes get cheaper, we are likely to buy more of them, but not many more, whereas fruit juice sales can be changed markedly by changes in price. This responsiveness to price is measured by the *price elasticity of demand*, and Table 3.1 shows how it varies over a range of food products.

There are two points to note from Table 3.1:

(1)   Basic foodstuffs generally have low price elasticities of demand. Most of us want to fill the potato-shaped hole inside ourselves at every meal so we need to buy basic foodstuffs no matter what the price, but once the hole is filled we won't need to buy more of the basics, no matter how cheap they are.

(2)   Conversely, the more a product might be termed a luxury, or the more substitutes it has, the more likely it is that the quantity demanded will be responsive to price changes. That's why the price elasticity of demand for carcase meat is greater than those for bread and potatoes.

Table 3.1 Estimates of price elasticity of demand for various food products.

Product	Price elasticity
Bread	−0.09
Fresh potatoes	−0.21
Processed potatoes and frozen chips	−0.29
Sugar and preserves	−0.24
Liquid milk	−0.29
Carcase meat	−1.37
Broiler chicken	−0.13
Frozen convenience meat and meat products	−0.94
Fresh green vegetables	−0.58
Processed fruit and fruit products	−1.05

Source: Ministry of Agriculture, Fisheries and Food (1989) *Household Food Consumption and Expenditure, 1989*, HMSO, London.

There's more on the effects of substitute price changes in the next section.

Thus, it appears that the price elasticity of demand for food products is generally low. It seems likely that the price elasticity of demand for food raw materials at the farm gate is lower still because some of the effects of price variation are damped by adjustments of processing and retailing margins. This low price elasticity has some important implications for farmers, as Figs 3.1 and 3.2 should demonstrate.

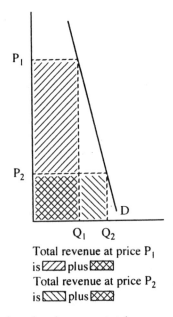

Total revenue at price P_1
is ▨ plus ▧

Total revenue at price P_2
is ▨ plus ▧

Fig. 3.1 The effect of a price change on total revenue for a price inelastic product.

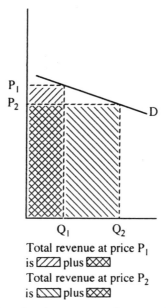

Total revenue at price P_1
is ▨ plus ▩
Total revenue at price P_2
is ▨ plus ▩

Fig. 3.2 The effect of a price change on total revenue for a price elastic product.

Fig. 3.1 demonstrates:

(1) That a small change in the quantity of a price inelastic product (Q_1 to Q_2) which is sold in a market can produce a big effect on its price (P_1 to P_2). This explains why short run supply changes brought about by weather and disease can produce *price instability* in many agricultural products. Fig. 3.2 shows that the opposite happens with a price elastic product.

(2) That a *decrease* in price from P_1 to P_2 in a price inelastic product does not produce a compensating increase in the quantity demanded, so the *total revenue* (price × quantity) received by the producer *goes down*. If you're not sure about this, remember that the area under the dashed price and quantity lines shows the total revenue. Thus, the total revenue when price is P_1 is greater than when the price is P_2 (and vice versa for a price elastic product, as Fig. 3.2 shows).

⇨ **Questions**_____

1 Using Table 3.1, calculate the effect that a 10% increase on the price of bread would have on the quantity of bread demanded? (If you can't remember the formula, look it up in the previous chapter.)

2 Again using Table 3.1, what is the size of the price change required to bring about a 10% increase in the demand for liquid milk?

3 It was once calculated that the price elasticity of demand for butter in the UK was –0.02, whereas the corresponding figure for UK-produced brands of butter was –0.16. Can you explain why the two figures should be different?

4 As a result of the drought in 1976, UK consumption of potatoes fell by about 20%, but prices rose by about 300%. Can you explain why this happened?

5 Would the total revenue received by UK potato growers have increased or decreased as a result of the drought in 1976?

3.3   Substitute price changes and demand

The price elasticities of demand for specific products are generally higher than those for the food groups to which they belong, as a result of substitution effects. For example, most people like to have a yellow fat to spread on bread and some are quite choosy about whether they spread butter or margarine, but most people find one brand of butter quite a good substitute for another. If the price of that brand goes up they might switch to another brand, but if the price of all butter and margarine goes up they are unlikely to eat dry bread.

When the price of lamb goes up, people will buy less of it and look round for a substitute. Beef is a pretty good substitute, so the quantity of beef demanded will rise. We can measure the extent to which this happens, you remember, by the *cross-price elasticity of demand*, and some examples of cross-price elasticities are shown in Table 3.2.

Table 3.2   Estimates of price and cross-price elasticities of demand.

	Elasticity with respect to the price of beef and veal
Beef and veal	–1.25
Mutton and lamb	0.36
Pork	0.08

Source: Ministry of Agriculture, Fisheries and Food (1986) *Household Food Consumption and Expenditure, 1989,* HMSO, London.

Table 3.2 shows that when the price of beef rises by 1% the quantity of beef demanded will fall by 1.25%, but 0.36% more lamb and 0.08% more pork will be demanded.

Obviously, the same kind of thing can happen with inputs too because one brand of concentrated feedingstuffs is a good substitute for another, just as one brand of fertiliser is a good substitute for another.

⇨ **Questions**_____

Now see if you can answer the following:

6 Would you expect the cross-price elasticity of demand for lamb and mint sauce to be positive or negative?

7 Would an increase in the price of butter shift the demand curve for margarine to the left or to the right? What effect would this shift have on the price of margarine, assuming that other things stayed the same?

3.4 Income changes and demand

The capacity of the human stomach is limited, although we may sometimes wish that it were not so. Consequently, when people have enough to eat, they will not spend extra income on greater quantities of food, although they may buy more expensive kinds of food (trading up, for example, from mince to steak). Thus, a tenfold increase in income will not produce a tenfold increase in the quantity of food consumed, as Ernst Engel (1821–96), Director of the Prussian Bureau of Statistics, argued in a paper published in 1857.

His conclusion is remembered as *Engel's Law*: as people get richer, the proportion of their income spent on food declines. This law has several important implications for the farming industry:

(1) It is the *proportion* of income spent on food that declines. Rich people spend more on food than poorer people, but they spend a smaller percentage of their income, as Table 3.3 shows.

Table 3.3 Income groups and food expenditure in 1994.

Gross weekly income of head of household (£)	Expenditure on food per person per week (£)
Over 560	15.68
290–559	13.37
140–289	12.58
Under 140	11.53

Source: MAFF (1994) *National Food Survey, 1994*, p. 19, HMSO, London.

(2) As the national income increases we would expect the percentage of total consumer expenditure going on food to decrease, and it does, as Table 3.4 shows.

Table 3.4 Percentage of total consumers' expenditure spent on.

Year	Household food	Meals out
1978–80	17.6	4.2
1985	14.1	4.3
1990	11.9	5.6
1992	11.9	5.5
1995	11.0	—

Source: MAFF (various years) *Agriculture in the United Kingdom*, table 1.1, HMSO, London.
(—) no data available.

(3) The income elasticity of demand for food in general in rich countries is low. The latest estimate for the UK is –0.01[1], which, if it's true, means that as the national income increases people will actually spend less on food. However, since the standard error calculated for this estimate is actually twice the size of the figure itself, we are justified in wondering whether this is really the case. In previous years estimates of income elasticity of demand have been around 0.2, which means that a 3% increase in national income (which is a pretty good annual rate of growth) would only produce an increase in the demand for food of (3 × 0.2 =) 0.6%.

(4) Income elasticities of demand for food are higher in low income countries.

There are also significant differences in *income elasticity* between one type of food and another. In general, we would expect the response to an increase in income to be positive but inelastic. In other words, if a person's income increases by 1%, we would expect them to spend a bit more on food, but not as much as 1% more. But we would also expect to find that their extra spending was concentrated more on things which are expensive, tasty, but not necessarily very filling, and less on basic stodge. If the income elasticity of demand for a particular food is negative, it implies that people actually spend *less* on it as they get richer.

Looking at Table 3.5, we can see if the figures indicate what we expect. They do indeed show that people do not buy more basic foods like bread and potatoes when their incomes increase, but that they buy more yoghurt, cheese, poultry, fruit and vegetables. The striking thing about Table 3.5 is how low the positive figures are (only fruit juice

Table 3.5 Estimates of income elasticity of expenditure for
household foods, 1989.

Product	Income elasticity
Liquid milk	−0.40
Yoghurt	0.58
Cheese	0.22
Processed cheese	−0.12
Carcase meat	−0.01
Beef and veal	0.08
Mutton and lamb	−0.21
Pork	−0.05
Bacon and ham, uncooked	−0.28
Broiler chicken, uncooked	−0.08
Other poultry, uncooked	0.24
Fish	−0.05
Eggs	−0.41
Butter	−0.04
Margarine	−0.44
Sugar and preserves	−0.54
Fresh potatoes	−0.48
Fresh green vegetables	0.13
Other fresh vegetables	0.35
Fresh fruit	0.48
Apples	0.32
Other fruit and fruit products	0.69
Fruit juices	0.94
Bread	−0.25
Breakfast cereals	0.01

Source: MAFF (1989) *Household Food Consumption and
Expenditure, 1989*, pp. 43–44, HMSO, London.

comes anywhere near 1) and how many of the figures are negative. We
might wonder if the really rich eat anything at all!

One explanation for the large number of negative figures in Table 3.5
might be changes in taste (which we shall examine in more detail in the
next section). Perhaps as people get richer they become more worried
about how fashionably slim they look (who was it that said 'You can
never be too rich or too thin'?), and their food buying habits reflect this.
If fat is a feminist issue, perhaps it's a farming issue too, because the
implication of Table 3.5 for agriculture, as we saw when we looked at
the income elasticity of expenditure for food as a whole, is that
increasing national income is not going to produce big increases in
demand. On the other hand, fruit and vegetable growers might expect
to do better than potato producers.

⇨   **Questions**_____

Now see if you can answer the following:

8 If national income increases by 3%, would you expect the quantity of apples demanded to increase or decrease, and by how much? (Use the data in Table 3.5.)

9 If consumers' incomes increase, will the demand curve for beef shift to the left or to the right? And by a lot, or only a little bit?

10 Would you expect the income elasticity of demand for bread to be the same in India as it is in the UK?

3.5   Changes in taste and demand

Once you have watched a small child eat baked beans and mint choc chip ice cream together out of the same bowl, you realise that taste in food is a complex business. It is affected by all sorts of sociological, psychological and anthropological variables. People in different social classes and from different ethnic backgrounds eat different kinds of food. We eat some foods for ceremonial purposes (Christmas dinners and birthday cakes). Sometimes we eat or don't eat because we feel happy or unhappy. But these kinds of variables change very slowly over time. In the medium term, two factors stand out as potentially affecting the demand for food: advertising and health considerations.

Advertising

Most advertising is brand advertising, and so relates to processed food. Thus, we might argue that it does not directly affect the agricultural industry. On the other hand, there are some examples of generic advertising (the sort that invites you to eat more meat or drink more milk without referring to the products of a specific firm) which have been produced by producer organisations such as the old Milk Marketing Board or the Meat and Livestock Commission. They do not seem to have been a major influence on demand.

Health

Various official reports into the health implications of diet have been produced in recent years, most of which have argued that saturated fats (such as those found in dairy products and red meats) may be associated with heart disease. They have probably had some effect, at least, on the consumption of dairy products and the rise in consumption of white meat at the expense of red meat. They have also argued that more dietary fibre is good for health, which may explain some of the relatively high income elasticities for fruit and vegetables that we noticed earlier.

⇨ **Questions**_____

11   Would a *successful* advertising campaign for the curiously named Stainless Steel Bulk Tank Yellow Dairy Fat Spread ('it comes from contented cows in flower-strewn meadows, and spreads straight from the fridge') shift the demand curve for the spread to the left or the right, and, if everything else stays the same, what effect would that have on its price?

12   (a) Would you advise the manufacturers of Stainless Steel Bulk Tank Yellow Dairy Fat Spread to find a new brand name, and, if so, why?
(b) Why do manufacturers of Stainless Steel Bulk Tank Spread base their adverts on the idea that 'it comes from contented cows in flower-strewn meadows'?

3.6   Population change and demand

A population increase means more mouths to feed, and so the demand for food rises. Generally, population growth rates in developed countries are low, as Table 3.6 shows, which means that the agricultural industry cannot rely on burgeoning populations to increase the demand for its product.

It is interesting to note, in comparison, that the annual change in the Turkish population is about 2.4%, and still higher figures can be found in some African and South American countries.

There are also more subtle changes which affect the relationship between population and food demand. Here are some examples:

* Older people eat different kinds of food from younger people, so as the population ages the pattern of demand changes.
* The number of single-person households is increasing, and they tend to eat more convenience foods.
* Manual workers tend to eat more calories than office workers, but there are fewer of them than in the past.
* Ethnic background affects the type of food demanded, so changing ethnic patterns have changed the demands for some kinds of food, e.g. there is now a market for some Chinese vegetables such as pak choi.

Table 3.6 Population change in developed countries.

	End of 1992 population (millions)	Average annual percentage change 1983–1992
Belgium	10.068	0.22
Denmark	5.181	0.13
Germany	80.614	comparison impossible
Greece	10.346	0.45
Spain	39.114	0.19
France	57.530	0.47
Ireland	3.560	0.07
Italy	56.933	−0.01
Luxembourg	0.395	0.79
Netherlands	15.239	0.57
Portugal	9.860	−0.23
UK	57.959	0.26
Sweden	8.692	0.42
Austria	7.910	0.47
Finland	5.055	0.33
USA	255.020	0.77
Canada	27.438	0.92
Japan	124.336	0.36

Source: Eurostat (1984, 1994) *Basic Statistics of the Community*, 24th and 31st edns., Table 3.1, Brussels.

3.7   Some long-term changes

The National Food Survey has now been carried out for over 40 years and some interesting changes have emerged over that time, as Table 3.7 shows. It is important to remember that all of the factors we have discussed work at the same time, so that the effect of one on the demand for food might be offset or reinforced by another. As people get richer you would expect them to eat more meat, but they are told that red meat contains saturated fats which have health implications, which might perhaps explain the rise in the consumption of chicken, especially when you remember that the real price of chicken is now less than it was 40 years ago.

Table 3.7 Household consumption of main food groups (ounces per person per week, except where otherwise stated).

Food group	Year		
	1950	1970	1994
Milk and cream (pints)	5.21	5.08	3.97
Cheese	2.54	3.59	3.74
Meat	29.85	39.53	33.29
Fish	6.62	5.35	5.12
Eggs (number)	3.50	4.66	1.86
Fats	11.61	11.95	7.98
Sugar and preserves	16.43	19.51	6.60
Vegetables	98.68	92.37	73.46
Fruit	18.09	25.52	34.17
Cereals	81.67	63.19	50.97
Beverages	2.72	3.61	2.22

Sources: MAFF (1991) *Household Food Consumption and Expenditure, 1990*, p. 93, HMSO, London; MAFF (1995) *National Food Survey, 1994*, p. 5, HMSO, London.

⇨ **Questions**_____

The two final questions are concerned with the material in the sections on population and long-term changes.

13 Will an increase in population shift the demand curve to the left or the right? And by a lot or a little bit?

14 Bringing together what you have discovered in this chapter, can you explain why the changes in food consumption between 1950 and 1970, and 1970 and 1990, shown in Table 3.7, have happened?

3.8 Summary

As far as the agricultural industry is concerned, several conclusions can be drawn from this study of the demand for food:

(1) The most important is that the demand for agricultural products is not likely to increase dramatically or quickly. In fact, the demand for farm products is pretty static. Populations in EU countries are not increasing very quickly and, although incomes are growing, the proportion of any extra income spent on food is small.

(2) Low price elasticities can produce price instability, and this means that producing more output will not necessarily increase the total revenue of the industry in a free market.

(3) People may be eating more exotic fruit and vegetables, but many of them are not produced within the EU and have to be imported.

(4) Other taste changes are produced by increasing health-consciousness which, if anything, seems to be decreasing the consumption of the cereals, meat and dairy products which form the staple products of farming in the northern countries of the EU.

(5) On the other hand, it is not impossible to envisage world demand rising in the long term if incomes in third world countries, which have high population growth rates, increase.

Note

1. MAFF (1989) *Household Food Consumption and Expenditure, 1989*, p. 43, HMSO, London.

Chapter 4
Changes in Agricultural Output

4.1 Introduction

One of the most remarkable features of agriculture in the second half of the twentieth century is the way in which its output has increased. This has not only happened in Britain but in most of the developed countries of the world. Paradoxically, while these output increases have provided more and cheaper food for many people, they have also produced problems: surplus products are expensive to store and can disrupt the world market if they are sold at less than their costs of production.

In Chapter 5 we shall examine the reasons for these increases in supply but before we do that it is important to find out more about what has actually happened.

4.2 The output of UK agriculture

The simplest way of illustrating output changes in agricultural output is to tabulate the tonnage of production (as in Table 4.1), although this approach is not without its problems.

Table 4.1 is quite good at illustrating the startling size of some of the changes in the output of British agriculture in the twentieth century: a sixfold increase in wheat production; a fourfold increase in milk production; a trebling of pig production; a trebling of poultrymeat production in only 30 years; and potato production doubled. At the same time, the story is clearly more complex than a simple increase in output. Oat production has declined steadily, which is not surprising in view of the replacement of horses by tractors. Sugar beet was just not produced to any measurable extent in 1905 and then expanded rapidly after its introduction as a result of government policy during the inter-war years. Equally, the output of wheat and barley might well have been higher in 1905 in the absence of competing imports and was stimulated by price support after 1939. Beef and barley output has declined in the

Table 4.1   The physical output of UK agriculture from 1905 to 1993 ('000s tons).

Product	Year				
	1905	1936–39	1959–60	1985	1993
Wheat	1 654	1 677	2 827	12 046	12 857
Barley	1 415	777	4 080	9 740	6 167
Oats	2 102	1 971	2 212	614	487
Potatoes	3 823	4 950	7 026	6 892	7 117
Sugar beet	—	2 784	5 598	7 717	8 550
Milk (m. litres)	3 517	7 105	10 488	15 408	14 085
Eggs (m. dozen)	118	541	1 069	1 059	800
Beef and veal (dcw)	652	587	759	1 124	877
Mutton and lamb	261	198	240	313	380
Pigmeat	273	442	704	945	1 023
Poultrymeat	40*	79**	262	874	1 080
Wool	31***	34	38	41	67

*   Estimated from 1936–39 production, roughly halved to take account of the change in poultry numbers.
**   Production plus imports minus exports, so probably exaggerated.
*** Estimated as roughly the same as in 1925, for which figures are available, since sheep numbers were roughly the same.
m. = million
dcw = dressed carcass weight
Sources: D. Grigg (1989) *English Agriculture: An Historical Perspective*, p. 6, Blackwell, Oxford. H.F. Marks and D.K. Britton (1989) *A Hundred Years of British Food and Farming: A Statistical Survey*, Taylor and Francis, London; MAFF (1994) *Agriculture in the UK, 1993*, HMSO, London.

last ten years while sheep production has continued to increase. Consequently, since we cannot add tons of barley to dozens of eggs, we must look for another way of demonstrating the *overall* change which has affected the agricultural industry.

4.3   Other methods of measuring output change in UK agriculture

Table 4.2 shows Grigg's estimates of the annual rates of increase in agricultural output for the twentieth century.

The important point that emerges very clearly from this table is the very rapid rate of increase in output in the post-World War II period. Grigg's estimate of the same growth rate for the period between 1780 and 1801, which is often identified as a period of 'agricultural revolution', is only 0.75% per annum.

Grigg's methodology for producing these estimates appears to be

Table 4.2   Average rate of increase in agricultural output, 1900–85.

Period	Rate of increase (% p.a.)
1900/4 to 1922/4	0.0
1922/4 to 1936/9	1.6
1936/9 to 1960	2.9
1960 to 1985	2.7

Source: D. Grigg (1989) *English Agriculture: An Historical Perspective*, p. 7, Blackwell, Oxford.

complex. A cruder method of solving the problem of adding together all the various outputs of the agricultural industry is to take the published figures for the gross output of the industry and adjust them to take account of changes in the value of money. The results of this exercise are shown in Table 4.3.

Table 4.3   Gross output of UK agriculture at current and constant (1990) prices.

Year	Gross output (£m) at current prices	Gross output (£m) at constant (1990) prices
1938	300	4918
1940	440	5176
1945	625	5342
1950	987	6130
1955	1396	7051
1960	1607	8878
1965	1979	10323
1970	2458	11097
1975	5012	11314
1980	9000	12596
1985	11996	13613
1990	13818	13818
1993	15484	14635

Sources: H.F. Marks and D.K. Britton (1989) *A Hundred Years of British Food and Farming: A Statistical Survey*, pp. 149, Taylor and Francis, London; MAFF (1994) *Agriculture in the UK, 1993*, HMSO, London. These figures for gross output in current prices are converted into constant prices using the agricultural price index (the figures are in Marks and Britton, p. 150, and the *Annual Abstract of Statistics*) re-based on 1990 = 100.

These figures suggest that the overall output of UK agriculture, after taking account of changes in the value of money, has risen almost threefold since the inter-war period.

4.4 Output changes in the EU

Similar figures to those in the UK are found in some, but not all, other European countries, as Table 4.4 indicates.

Table 4.4   Wheat production ('000s tons) in various countries.

Year	France	Germany	Italy	Netherlands	Spain
1905	9 110	4 187	4 682	128	2 518
1938	9 800	6 250	8 184	434	2 870*
1960	11 010	6 421	6 794	590	3 520
1970	12 921	7 794	9 689	643	4 060
1985	28 871	9 866	8 516	851	5 326
1993	28 368	15 720	4 096	1 035	4 205

* 1939 figures.
Sources: B.R. Mitchell (1978) *European Historical Statistics 1750–1970*, table C2, Macmillan, London; Eurostat (various editions) *Agriculture: Statistical Yearbook*, Brussels.

This attempt to illustrate output changes is even more limited than Table 4.1, although it does have the merit of showing developments over a long period of time. An alternative approach is to use *indices*, which express total output as a percentage of the output in a base year or period, as in Table 4.5. Section 4.6 covers indices in more detail.

Table 4.5   Volume index numbers of final agricultural output, EU-12* (1984–6 = 100).

1975	80.8
1980	92.0
1985	99.4
1990	105.5
1993	106.8

* Pre-1983 data for EC-10, rebased from indices using 1975 and 1980 as base years.
Source: Eurostat (various editions) *Agriculture: Statistical Yearbook*, Brussels.

4.5 World output

Just as UK and European output of agricultural products has grown so has world output, and at roughly similar rates, as Table 4.6 shows.

However, in cereal production, the picture is more complex as Table 4.7 shows, with the growth rate in output decreasing during the 1980s in

Table 4.6   Index numbers of world agricultural output (1979–81 = 100).

1948–52	44.9
1955	53.2
1960	61.9
1965	69.4
1970	79.4
1975	89.2
1980	99.1
1985	114.2
1990	125.9
1994	129.4

Source: FAO (various editions) *Yearbook of Agricultural Production*, FAO, Rome.

Table 4.7   The growth in world cereal production 1965–91: tonnages and growth rates.

	World cereal production (million tonnes)			Growth rate (% p.a.)	
	1965–70	1980–85	1991	1965–80	1980–90
Industrialised countries	411.6	606.8	612.5	2.6	0.2
Former Soviet Union	150.9	172.0	164.9	2.3	2.4
Eastern Europe	50.7	76.3	81.0	2.9	1.2
Developing countries	418.9	694.2	837.4	3.4	2.5
World	1 032.1	1 549.2	1 695.8	3.1	1.9

Source: OECD (1993) *World Cereal Trade: What Role for Developing Countries?*, p. 41, OECD, Paris.

industrialised countries. This is presumably a result of the trade problems of that decade, although whether the same explanation can be advanced for the smaller decrease in the output of developing countries is less certain.

The conclusions from all these figures are pretty clear. With a few exceptions, output is increasing, especially in industrialised countries, at a rate which is very rapid in comparison with the growth in farm output in previous periods. Chapter 5 will examine the reasons for this increase.

4.6   Comparisons over time: index numbers and inflation

In this chapter we have used various measures of change in agricultural output. Table 4.1 shows how the physical output of UK agriculture has changed in the twentieth century, and we can see that the output of most commodities has increased more or less continuously up to 1985. There are several questions that we can ask of these figures, and several ways of answering them.

Measuring changes: the use of index numbers

The first question might be 'Which experienced the greater increase in output between 1936–39 and 1959–60: wheat or potatoes?'. Here's one way of working it out (all figures in '000s tons, from Table 4.1).

	Wheat output	**Potato output**
1959–60	2827	7026
minus		
1936–39	1677	4950
Difference	1150	2076

At first sight, therefore, potato output increased more than wheat output. But obviously, potatoes started at a higher figure, so we can get a fairer comparison of the *relative* increase by measuring the percentage change in the output of the two commodities, or producing a simple index number, using the formula below.

$$\text{Index number for the second period} = \frac{\text{Second period data}}{\text{First period data}} \times 100$$

For the wheat and potato example, this looks like:

	Wheat output	**Potato output**
1959–60	2827	7026
1936–39	1677	4950

So,

$$\text{Index number} = \frac{2827}{1677} \times 100 \qquad \frac{7026}{4950} \times 100$$

$$= 168.6 \qquad\qquad 141.9$$

In effect, we have expressed the figure for the later period as a percentage of the figure for the first period, and we can see that the relative increase in wheat output was greater than in potato output. And, of course, we could also express the data for 1985 and 1993 in Table 4.1 as percentages of the 1936–39 data and extend the series. This is a very useful technique for comparative purposes, so it is worth learning how to produce simple index numbers. Here is a set of simple rules to follow:

(1) Decide on a base year, which will take the value 100 (written 'year = 100').

(2) List the individual years in the series.

(3) Work out the index numbers for the individual years using the formula

$$\text{Index number for individual year} = \frac{\text{Data value for individual year}}{\text{Data value for base year}} \times 100$$

Using these rules, and some of the data from Table 4.4, we can convert the figures for wheat production in the Netherlands into index numbers:

(1) Let us take 1970 as the base year (1970 = 100)

(2) List the individual years in the series:

Wheat production in the Netherlands

Year	Production ('000s tons)
1905	128
1938	434
1960	590
1970	643
1985	851
1985	1017

(3) The index number for 1905 is $\dfrac{128}{643} \times 100 = 19.91$

and for 1938 : $\dfrac{434}{643} \times 100 = 67.50$

and so on, to give us the following:

Wheat production in the Netherlands

Year	Production ('000s tons)	Index (1970 = 100)
1905	128	19.91
1938	434	67.50
1960	590	91.76
1970	643	100.00
1985	851	132.35
1992	1017	158.16

⇨ **Questions**_____

1   Now see if you can convert the wheat production figures for France and
    Germany in Table 4.4 into index numbers using 1985 as the base year,
    and decide whether wheat production increased more in France or
    Germany between 1905 and 1992. Here are the figures:

Wheat production ('000's tons)

Year	France	Germany
1905	9 110	4 187
1938	9 800	6 250
1960	11 010	6 421
1970	12 921	7 794
1985	28 871	9 866
1992	30 613	15 472

You have now produced some simple index numbers and used them
to answer a question about some statistical data. Using the same basic
technique, you can also investigate changes in more complicated sets of
figures. You will have seen that Table 4.5 contains index numbers of
the volume of all agricultural output of the whole EU for the period
1975–92. In other words, it sums up the data for several countries and
many different commodities. How is it done?

If we go back to the data in Table 4.4, we can produce a simple
example to illustrate the process, via the following steps:

Step 1: Take the figures in a row of data, and average them. So if we
take the figures for 1905 we get:

$$\frac{(9110 + 4187 + 4862 + 128 + 2518)}{5} = 4161$$

Step 2: Do the same for all the other rows of data in the table. This
gives us the following figures:

1905	4 161.0
1938	5 507.6
1960	5 667.0
1970	7 021.4
1985	10 686.0
1992	10 958.0

Step 3: Treat one year as the base year and use it to express the other
figures as an index, just as we did with the data for individual countries

when constructing a simple index. So if we take 1970 as the base year, the index for 1905 will be:

$$\frac{4161}{7021.4} \times 100 = 59.3$$

And for the other years it will be:

1905	59.3
1938	78.4
1960	80.7
1970	100.0
1985	152.2
1992	156.1

Thus, we have now constructed an index for wheat production in five EU countries for various years in the twentieth century to the base 1970 = 100. Only the calculations for 1905 are shown, so try working through the calculations for the other years until you are sure that you know what is going on.

Straightforward though it is, this is a potentially very powerful technique, because we can use it to construct indices which bring together large amounts of data into a single column of figures. We can also use more complex methods of combining the data.

Suppose we were looking at prices, for example. We might look at price changes in a collection or basket of goods, find the average price change, and express it as an index. But suppose we found that the price of food, which accounts for nearly one fifth of all our expenditure, had gone down a bit, whereas the price of pencils, which accounts for a tiny proportion of our expenditure, had gone up a lot. What would have happened to the cost of living? It's obvious that food prices should be much more influential than pencil prices, but in a simple index they will have equal weight. The solution to this problem is to produce a *weighted index*, in which the individual price changes are weighted by the proportion of expenditure they account for, or their contribution to total output, or whatever seems most appropriate.

The retail price index (RPI), which is commonly quoted in the press as the measure of inflation, is an example of just such a weighted index (Table 4.8). We won't bother to go into the details of how the weights are incorporated into the calculations, because we are more interested in using the RPI, as we will in the next section on how to deal with inflation, than in constructing it[1].

Table 4.8   The Retail Price Index (1985 = 100).

Year	Index	Year	Index	Year	Index
1948	8.4	1964	15.0	1980	70.7
1949	8.6	1965	15.7	1981	79.1
1950	8.8	1966	16.3	1982	85.9
1951	9.6	1967	16.7	1983	89.8
1952	10.6	1968	17.5	1984	94.3
1953	10.9	1969	18.4	1985	100.0
1954	11.1	1970	19.6	1986	103.4
1955	11.6	1971	21.4	1987	107.7
1956	12.1	1972	23.0	1988	113.0
1957	12.6	1973	25.1	1989	121.8
1958	13.0	1974	29.5	1990	133.3
1959	13.0	1975	36.1	1991	141.1
1960	13.2	1976	42.1	1992	146.4
1961	13.6	1977	48.8	1993	148.7
1962	14.2	1978	52.8	1994	152.4
1963	14.5	1979	59.9	1995	157.6

Sources: CSO (1995) *Economic Trends, Annual Supplement 1995*, p. 148, HMSO, London; 1994 and 1995 figures estimated from data in CSO (1996) *Monthly Digest of Statistics*, HMSO, London.

Dealing with inflation

At the beginning of this section we referred to Table 4.1, in which we could see that the output of most commodities had increased more or less continuously up to 1985. But this is not the case for oats and potatoes. So how can we decide whether or not the *overall* output of the industry has changed? After all, we can't add tons of wheat to litres of milk and dozens of eggs and get a sensible answer. You might now say 'Ah yes, but we could convert them all to indices', and so we could. But there is another approach too, which has the additional advantage that it also takes into account the value of the products. It involves multiplying the physical output by the price at which it sold at the time it was produced. This approach puts everything in monetary terms, so that we *can* add the output of wheat to the output of milk and eggs because they are now all the same units. And it takes account of the fact that a ton of wheat is worth more than a ton of potatoes or sugar beet. The results of doing this for every fifth year in the period 1938–93 are shown in Table 4.3.

Many of us have spent much of our lives in a period which is, historically, one of relatively high inflation so we are familiar with the idea that the value of money can change. The paperback novel for which you paid £4.99 a couple of years ago might now cost £5.99. When Penguin books first appeared in 1935 they cost sixpence (2.5p in modern decimal coinage), but they were no thinner or less well written.

So, obviously, comparing the output of agriculture in 1938 prices with its output measured in 1993 prices is not comparing like with like. This is why Table 4.3 contains two columns. One gives the gross output in *current* prices, the ones used at the time when the output was produced. The other gives it in *constant* prices: all the output figures are re-calculated to take account of inflation using 1990 as the base year and an index of the price of agricultural products.

Converting current prices to constant prices is a matter of simple arithmetic if you have a table of values for the RPI, which you do in Table 4.8. It is based on 1985 = 100, so using it enables you to convert all prices to 1985 values. For example, take the following figures for wheat prices (the average price for the year, in £ per tonne, given in MAFF (1989, 1991) *Agriculture in the UK*, HMSO, London).

1985	112.2
1988	105.1
1991	116.6

Do these figures imply that the price of wheat fell and then rose again? Clearly they do, in current price terms. But what about the purchasing power of the money that a farmer receives from selling a tonne of wheat? In order to work that out we need to convert these current prices into constant prices.

Step 1: Find the RPI for these years from Table 4.8. The values are:

1985	100.0
1988	113.0
1991	141.1

Step 2: Insert the figures in the formula.

$$\frac{\text{Constant price}}{\text{for the year}} = \frac{\text{Current price}}{\text{for the year}} \times \frac{\text{Base year RPI} \, (= 100)}{\text{RPI for the year}}$$

So obviously, for 1985, the base year, the current and constant prices will be the same. For 1988 the constant price will be:

$$105.1 \times \frac{100}{113.0} = 93.01$$

And for 1991:

$$116.6 \times \frac{100}{141.1} = 82.64$$

Step 3: Tabulate the constant prices.

1985	112.2
1988	93.01
1991	82.64

Thus, it appears that the price of wheat, after allowing for inflation (or, as is sometimes said, the price in *real* terms), fell steadily over this period, despite what you might have thought when you first examined the current prices.

⇨ **Question** _____

Now practise this technique using the following example, which will tell you whether the price of hiring a car has risen or fallen since 1957.

2   Taking either a small car (Minor or Metro) or a large car (Cambridge or Rover 416), did the real cost of car hire rise or fall between 1957 and 1994?

Cost per day's hire (£)			
From Blue Star Garages in 1957		From Avis Car Rentals in 1994	
Morris Minor	1.375	Rover Metro	41.0
Austin Cambridge	1.375	Rover 416	57.0

Usually, you will find it best to choose the RPI when looking for a measure of inflation with which to convert a current price series into constant prices. In some circumstances there are other indices which are used (as in Table 4.3) but, in general, think of using the RPI unless you have good reasons for doing otherwise.

4.7   Summary

- Output changes can be measured in tonnages, pounds sterling, or by using indices.
- UK agricultural output has risen almost threefold since the inter-war period.
- Output has also increased in the EU and the rest of the world.
- Index numbers can be used to make large amounts of data intelligible, and to take account of inflation.

Note

1.   If you would like to know more about the calculation of RPI, read Chapter 5 of C. Johnson (1988) *Measuring the Economy*, Penguin, Harmondsworth, which is clearly written and informative.

Chapter 5
The Supply of Farm Products

5.1 Introduction

We have seen that the output of agriculture has increased significantly over the twentieth century, and rapidly since 1950. This chapter attempts to explain why that has happened and what determines the supply of agricultural products.

In common with many other governments, the European Union, through its Common Agricultural Policy, intervenes in agricultural markets to manage supplies. Initially, the emphasis was on increasing output; since about the middle of the 1980s, and especially since the MacSharry reforms of 1992, supply control through quotas and set-aside schemes has been given greater prominence. In each case, policy makers need to know what affects the supply of agricultural products.

5.2 Economic theory and supply

Chapter 2, on the way markets work, contained a list of factors affecting supply (can you remember what they were?).

It is important to remember that these factors determine what the producer *intends* to supply; *actual* supply at any point in time may differ from intended supply as a result of the following.

- *The effects of weather and disease on crops and animals:* Sometimes the combination of sun and rain is ideal and the corn shoots up, the lambs and calves forget to die suddenly, sleepy sows lie on straw and not piglets, and production exceeds expectations. Then it's back to soggy sowing times, drought in May and June and south-westerly gales lashing August rains over the shrivelled crops while the combines rust idly in the corner of the field[1].
- *Time lags in production:* Higher prices may encourage farmers to produce more, but they have to wait for crops and animals to grow

before they can be sold. Pig production can be expanded faster than beef production simply because pigs have a shorter gestation and fattening period than cattle. On the other hand, if they are kept indoors, the problem of finding accommodation for them may offset this advantage.

• *Government policy:* Quotas and set-aside arrangements may mean that the farmer cannot produce the quantity which would be produced in their absence.

Having taken all that into account, the economist's theory of supply suggests that the *individual* farm's supply of a product will be determined by:

• The price of the product
• The price of other products which the farm might produce
• Costs of production
• The state of technology
• The objectives of the farmer.

The supply produced by the agricultural *industry* as a whole will be determined by the sum of the supply decisions of individual farmers, together with:

• The number of farms in the industry and whether they are big or small

5.3   Supply and the price of the product

There are two main issues to consider here: the relationship between supply and price, and the elasticity of supply and the factors which affect it.

We saw previously that higher prices encourage producers to supply more to the market, as the supply curve(S) in Fig. 5.1 indicates.

In the short term, when prices are high, it is worth using more feed or more fertiliser to produce more output.

When prices are low, it is harder to cover input costs. For example, if the costs of seed, fertiliser and sprays for a winter wheat crop are £220/ha and the yield is only 5 tonnes/ha, the crop needs to be sold for £44/tonne to cover these *variable costs*. Variable costs are those like seed, fertiliser and sprays which vary as output changes in the short term. To take another example, the cost of fuel required for a combine to cut a light crop will be less than the fuel cost in a heavy crop. Thus, fuel cost varies with output from harvest to harvest, so it is a variable cost. But you can't sack the combine driver just because it doesn't take long to

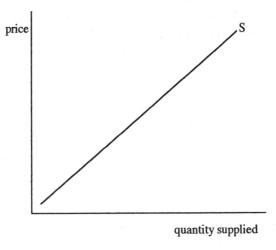

Fig. 5.1   The supply curve. S = supply.

cut the corn, so *regular* labour is a *fixed cost*. The amount of casual labour you require, however, would vary with the output (think about potato pickers) so casual labour is a variable cost.

In the long term, when prices are high, it is worth spending money on land drainage and other forms of reclamation, so that, for example, land which might otherwise only be rough grazing can produce more grass, and grazing land might produce arable crops.

When prices are low, production tends to retreat from these marginal lands, which is why it is possible to see the marks of cultivation on grass-covered hillsides all over the uplands; when prices were high (usually in wartime) they were worth cultivating but now they are not. Over a long period, fixed costs *can* vary; land drainage and moorland reclamation are good examples.

The *price elasticity* of supply – the extent to which the supply changes for a given change in price – depends upon four variables:

(1) *Time:* It takes time for crops to grow and animals to get fat or give milk, so the longer the growing period, the longer it will take for supply to respond to a price change. Thus, orchard trees, and especially forest trees, have very low price elasticities of supply.

(2) *Cost structure:* In the long term, it is not only necessary to cover the costs of seed, fertiliser and sprays – the variable costs – but fixed costs as well (labour, machinery, rent and other overheads). But since it may take some time to dispose of land or change labour costs, it may still be worth continuing to produce in the short term as long as variable costs are covered. Fixed costs in agriculture may form a big proportion of the total costs, which is

why supply often responds slowly to a reduction in price. The replacement of labour by buildings, and hired labour by family labour, accentuates this tendency.

(3) *Asset fixity:* Another reason why falling prices do not produce rapid reductions in supply is the low resale price of farm buildings and machinery. A potato harvester is a specialised bit of kit, and if potato prices go down other producers are unlikely to want to buy one, and only other producers would have any use for one. Thus, farmers feel that they might as well continue to produce as long as the returns cover the variable costs, and so supply fails to respond to falling prices. This phenomenon is known as *asset fixity*, and is brought to an end if the asset wears out while prices are still low.

(4) *Supply control:* In the form of quotas and set-aside, this prevents supply increasing in the event of prices rising, and even if quotas are transferable the cost of buying or leasing them will presumably dampen the effect of any price increase.

Thus, for most farm products there are good reasons for supposing that the response to falling prices will be sluggish, although the response to price increases may be greater as long as it is not constrained by some form of supply control. This is *not* the same as saying that there is a perverse supply response in agriculture, i.e. that when price decreases the quantity supplied *increases*. There is little evidence for this ever happening, especially over the long term.

⇨ **Questions**_____

Now use these questions to review your understanding of the material covered in the last section.

1 An arable farm has fixed costs of £565/ha. Its variable cost of producing winter wheat is £220/ha. It produces 7.25 tonnes/ha of feed wheat which sells at £90/tonne. It also receives Arable Area Payments of £250/ha. If these Area Payments are removed entirely, should it continue to produce wheat, or cease production in the long term or the short term?

2 If the average all pigs price (AAPP) decreases, will the quantity of fat pigs supplied to the market increase or decrease?

3 What are the main fixed costs likely to be on a hill farm?

5.4 Supply and the price of other products

Many agricultural politicians have found, to their dismay, that intervention in the market for one product can affect the market for other products.

Complementary products

These are produced together, so that changing the price of one affects the output of the other. Milk and beef, lamb and wool, and cereals and straw are obvious examples. Thus, when milk quotas were introduced the supply of beef was at first increased as more cull cows came on the market, and then reduced as there were fewer cows to have calves to sell into the beef trade. However, it is important to recognise that one of the products usually has a bigger influence on the production decisions than the other: nobody would keep more sheep because the price of wool, which only accounts for about 4% of the total output of a lowland fat lamb producer's output, had risen (although the position might be reversed in Australia).

Break crops are also complementary products. Since a break crop on a cereal farm helps to produce a higher output of grain, their production might continue even though the return they produced was very low. Thus, a price change in a break crop might not have a very big effect on the quantity supplied.

Competitive products

These use the same inputs: wheat and barley, sugar beet and potatoes, and beef and sheep are obvious examples. Consequently, if the price of one increases, more of it will be supplied, while less of its competing product is supplied. Remember that changing the price of another product which the farm produces or might produce will *shift* the supply curve to the left or right.

⇨ **Questions**_____

Now see if you can answer the following:

4   After the introduction of milk quotas in 1984, what would you have expected to happen to the supply curve for beef in the short term and then in the long term? Assuming no change in the demand curve for beef, what effect would this have on the beef price?
5   If the price of barley increases, will the supply curve for wheat shift to the right or left?

5.5   Supply and the costs of production

An increase in production costs will shift the supply curve to the left, and a decrease will shift it to the right, but some cost changes will be more influential than others. For breeding and rearing pigs, an increase in feed costs, which account for 86% of variable costs, will have a bigger effect than an increase in vet and medical costs, which only account for about 5% of variable costs[2].

Over the long term, as agriculture develops, farmers seem to buy more and more of their inputs from outside the farming industry. Before 1850, for example, UK farmers largely relied on farmyard manure for crop nutrients, farm labour for weed control, and hay and roots for winter feed. In 1995, fertilisers, pesticides and feedingstuffs account for 36.3% of the total costs (including purchased inputs, depreciation, interest, rent and hired labour costs) of UK agriculture. But in other EU countries, such as Greece and Portugal, farmers are less reliant on purchased inputs, which probably means that the price elasticity of supply is lower than it is in the UK.

⇨ **Questions**_____

You should now be able to answer the following:

6   If the cost of fertiliser rises, will the supply curve for cereals shift to the left or the right?

7   Which would have the greater effect on the supply of milk: an increase in the cost of artificial insemination, or an increase in the cost of purchased feedingstuffs?

8   Which would you expect to have the greater proportion of bought-in inputs: a hill livestock farm or a lowland arable farm?

5.6   Supply and the state of technology

Technical change has been one of the most important influences on the supply of agricultural products in the second half of the twentieth century. There have been new varieties of crops, new breeds of livestock, new pesticide and fertiliser formulations, new types of machinery, more sophisticated feedingstuffs, and many other examples of new technology. Whether it has been a good or a bad thing is a complicated question, but before you think about that you need to work out what brings it about and what effect it has on the supply curve.

What produces technical change?

We are so used to rapid technical change in agriculture that it is tempting to think of it as a natural process: researchers always come up with new ideas, and farmers always adopt them. In fact, while that may seem to have been the position in recent years, it has not always been so, and we have to explain why this apparently natural process occurs.

There are in fact two parts to the question:

(1) What produces innovation?
(2) What makes farmers adopt innovations?

This is because an innovation will have little effect if it is not adopted by farmers, and some are more quickly adopted than others.

The process of innovation

Innovation seems to be produced by several factors:

- Changes in relative prices of inputs: If labour gets relatively more expensive, it provides an incentive for people to try to invent ways of getting the job done with less labour, or making the same amount of labour more productive. The same might apply to land. If there is plenty of it, it doesn't matter if each hectare doesn't produce very much, but if the land price is high, or the farmer has only a small area on which to operate, there is an incentive to make each hectare produce a bit more. This is called *induced innovation*.
- Changes in output prices: This is a bit more difficult, because there are logical reasons for seeing two sides of the argument. If prices rise, farmers have an incentive to produce more, and so will be interested in innovations which enable them to do so. If prices fall, they will look for ways of maintaining their incomes, and will be interested in innovations which enable them to do that. Perhaps the answer is therefore that there will be different types of innovation in different circumstances: output increasing innovations, such as new varieties and pesticides, when prices are rising, and more interest in new products when the price of existing products falls. On the other hand, higher yielding varieties are seldom unwelcome in any circumstances.
- Funding of research and development: This can come from governments and ancillary industries. Few farms are big enough to have their own research departments, in the same way that a fertiliser or pesticide manufacturer might carry out research. (Those that are large enough often seem to be concerned with animal breeding, from Robert Bakewell of Dishley Grange in Leicestershire, who produced the New Leicester sheep and the Longhorn cattle in the eighteenth century, to Brian Cadzow of West Lothian in the 1960s and 1970s, who produced the Cadzow Improver ram.) Consequently, they are dependent on the activities of outside research organisations. Some of the first agricultural experimental

stations were established in Germany in the middle of the 19th century by organisations of farmers and landowners. At the same time in Britain, Lawes and Gilbert set up the Rothamsted experimental station, funded by the profits of Lawes' fertiliser firm, and a little later the Royal Agricultural Society also set up an experimental farm. But although they had some successes, agricultural science in Britain developed much faster after the university departments of agriculture were established, mostly in the 1890s, and the network of government-funded research stations came into being in the early years of the twentieth century. Then, later on, from the 1930s, firms in ancillary industries began to carry out their own product-development research. Government research funding may also be affected by product prices. It was noticeable that there was much discussion of the desirability of agricultural research in the late 1980s and early 1990s, when there was concern over surplus production.

Adoption of innovations

Some factors which determine an innovation's adoption are concerned with the innovation itself.

- What you don't know about doesn't exist, so the more it is discussed by the press, TV, radio, and the advisory services, and seen at agricultural shows and demonstrations, the more likely it is to be adopted. Indeed, the original purpose of agricultural shows was to promote the adoption of innovations.
- The more it costs to introduce it, the longer the adoption process will take, so new seed varieties are rapidly adopted, whereas something that requires new machinery and buildings, such as the replacement of hay by silage, can take much longer. Silage was first introduced into Britain in the 1880s, but was not widely adopted for nearly a hundred years.
- The more uncertainty there is about the effects of the innovation, and the more management expertise required to make it work, the less likely it is to be adopted. It used to be said that forward creep grazing of lambs was an excellent system for really good grassland managers, but it was never widely adopted because not enough farmers were good enough grassland managers.

Some are concerned with the characteristics of the adopting farmers.

- Their personal characteristics: Young, rich, well educated farmers running big and/or specialised farms are more likely to adopt an innovation than old, poor, farmers on small mixed farms.

- Their social relationships: Cosmopolitan risk takers and opinion leaders are more likely to adopt an innovation than socially isolated risk-avoiders.
- Their communication behaviour: Those who interact with other innovators and are capable of using the scientific literature are more likely to adopt an innovation than those who get most of their information from friends and neighbours and make little use of the mass media.

What are the effects of technical change?

Most innovations are output-increasing. There are some exceptions to this generalisation: mechanised fruit-picking may result in lower yields but may be adopted because it lowers costs, and hand-hoeing may be as effective at removing weeds as herbicides, but costs more. However, new breeds and varieties, fertilisers and feedingstuffs are adopted because they enable output to be increased from the same land area. Table 5.1 shows the extent to which yields have been increased in the second half of the twentieth century.

Table 5.1 Yields in the UK (tonnes per ha., except for milk in litres per cow).

Period	Wheat	Barley	Oats	Potatoes	Milk
Pre-war	2.31	2.09	2.05	16.8	2457
Early 1950s	2.81	2.62	2.35	19.9	2787
Early 1960s	3.96	3.45	2.76	22.2	3576
Early 1970s	4.24	4.41	3.73	29.2	3737
1980	5.88	4.43	4.07	34.5	4714
1990	6.97	5.21	4.96	36.4	5148
1995	7.66	5.75	5.46	38.3	5366

Sources: H.F. Marks and D.K. Britton (1989) *A Hundred Years of British Food and Farming: A Statistical Survey*, Taylor and Francis, London; MAFF (various editions) *Agriculture in the UK*, HMSO, London. Obviously the 'yields' of pigs and grazing livestock cannot be measured in the same way, but there is no reason to suppose that they have not increased to the same extent as those of the enterprises shown above.

Therefore, the overall effect of innovations is to move the supply curve of the individual farm to the right, from S_1 to S_2, as in Fig. 5.2.

Since there are so many farmers in the agricultural industry, no individual has much impact on the output of the industry as a whole, so an individual innovator will have no effect on the price. He will get the reigning market price whether he adopts the innovation or not and, since an output-increasing innovation shifting his supply curve from S_1

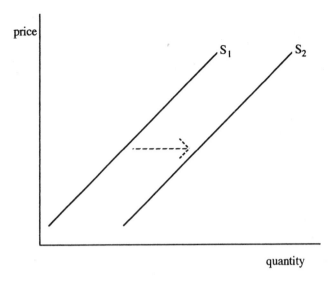

Fig. 5.2 The effect of technical change on the supply curve.

to S_2 will increase his total revenue (i.e. price received × quantity produced), he will gain by adopting the innovation (Fig. 5.3).

However, if many farmers adopt the innovation, the supply curve *for the industry as a whole* will shift to the right (S_1 to S_2), and since the industry demand curve is downward-sloping, the price will fall as supply increases. Indeed, since the price elasticity of demand is often

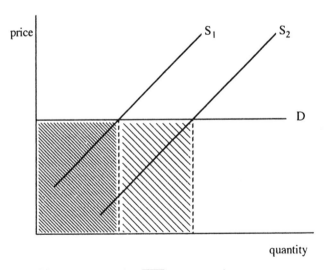

The area shaded thus ▨ represents the extra revenue for the individual

Fig. 5.3 The supply curve shifting to the right with a horizontal demand curve.

low, the new total revenue may be lower than the initial total revenue (see Fig. 5.4).

This is the relationship which led the American agricultural economist Willard W. Cochrane to write in 1958 about a 'treadmill with respect to technological advance'. In a free market, in which there is no government intervention, it makes sense for the individual farmer to adopt output-increasing innovations which reduce the prices and often the incomes received by farmers as a whole. Farmers are not, of course, unaware of this paradox. Some have suggested voluntarily withholding produce from the market in order to drive up the price, but not only is this often illegal unless part of an official marketing scheme, it is also ineffective, because there will usually be a few producers who refuse to participate. It is also important to remember that the food consumer benefits from lower prices and more plentiful supplies.

In the EU (and in other developed countries too), agricultural policy prevents the decrease in price which would occur in the unsupported market (assumed in Fig. 5.4). This means that the price does not fall as the quantity demanded increases. Therefore, the demand curve is horizontal. In this case, the agricultural industry as a whole receives the supported price, and the effect of the innovation is not to reduce the price but to increase the total revenue of the whole industry, as Fig. 5.5 demonstrates.

The consumer does not benefit from lower prices. The additional supply replaces imports if the EU is not self-sufficient in the commodity and even when it is, the excess goes into intervention stores at a cost to

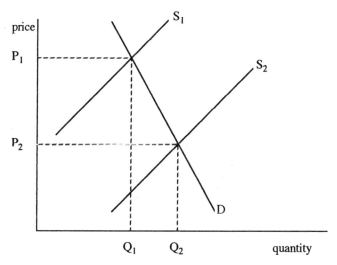

Fig. 5.4   The supply curve shifting to the right with a downward-sloping demand curve.

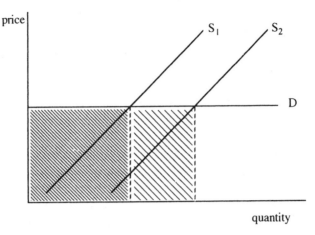

The area shaded thus ⬚ represents the extra revenue
for the whole industry

Fig. 5.5   The supply curve shifting to the right with the whole industry demand
curve horizontal.

the taxpayer. Since the supported prices, in real terms, may not be
maintained at their original level, the treadmill effect may still operate,
although not to the same extent as it would in an unsupported market.

⇨ **Questions**_____

You should now be ready to answer the following:

9   Can you explain the following figures without arguing for a backward-
sloping supply curve?

	UK milk output (million litres)	UK milk price (pence per litre, constant 1985 prices)
1965	10 710	21.8
1975	13 167	17.5
1985	15 242	14.5

10   If the price elasticity of demand for farm products is elastic (as it might
be in third world countries), what would be the effect of technical
change on the total revenue of the agricultural industry?

5.7 Supply and the objectives of farmers

Governments intervening in agricultural markets are likely to be interested in farmers' attitudes to risk. If farmers are risk-averse, any sort of price guarantees may increase supplies without prices changing at all.

The objectives of farmers are unlikely to change in the short term, but over the long term individual farmers may experience changes. If long-term price falls are predicted, for example, some farmers may decide that minimisation of bank borrowing is preferable to output maximisation. It also seems likely that the objectives of part-time farmers (whose numbers are increasing) will be different from those of full-time farmers. However, in the analysis of supply response it is still usually reasonable to go along with the assumption underlying supply theory, which is that the objective of the producer is to maximise profits.

5.8 Factors affecting the industry supply curve

So far, all the variables discussed have been those which affect the supply response of the individual producer. The supply from the whole agricultural industry is the sum of those individual responses, but there are some additional complicating factors:

- *The number of farms:* When prices fall, and marginal producers leave the industry because they can no longer make sufficient profit, the land they farm may be taken over by one or more of the remaining farmers, so the industry supply need not necessarily decrease. The rate at which land moves out of farming depends more on the demand for land for housing and forestry, as we shall see in Chapter 6.
- *The size of farms:* Small farms need to produce more intensively than big ones to produce the same level of profit, so an agricultural industry with many small farms will produce a different level of output from one with many big farms under similar conditions of price and cost.

5.9   Summary

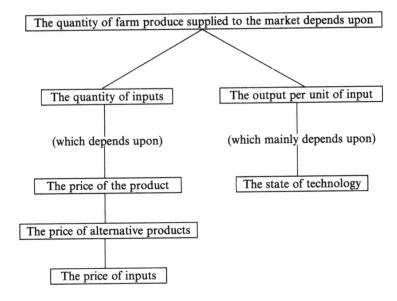

The main conclusions from this chapter are that the supply of farm products

(1)   is not very responsive to price changes; and
(2)   increases in the long term as a result of technical change.

Notes

1.   If you want to know more about the effects of weather and disease in the past, read E.L. Jones (1964) *Seasons and Prices*, George Allen and Unwin, London, or J.M. Stratton (1978) *Agricultural Records AD 220–1977*, John Baker, London.
2.   Calculated from J. Nix (1995) *Farm Management Pocketbook*, p. 82, Wye College Press, Ashford.

Chapter 6
Agricultural Inputs

6.1 Introduction

Economists conventionally divide the resources or inputs available to a business into four types:

- Land (the reward to which is rent)
- Labour (rewarded by wages and salaries)
- Capital (rewarded by interest)
- Management or entrepreneurship, i.e. the willingness to organise other inputs and take risks (the reward for which is profit).

Much economic theory treats these resources separately, as if they were provided by different people. In a multinational company or even on a large rented farm this may well be the case. But on a small, owner-occupied farm all four resources are provided by the same person: the farmer, who:

- Owns the land and the capital (the buildings, machinery and breeding stock).
- Provides the manual labour required to produce the output and the mental labour required to do the paperwork.
- Decides what to do and when to do it, and takes the risk that at the end of it all no profit will be made.

This means that the question 'what is the income of the agricultural industry?' is not always an easy one to answer, because farmers may receive rewards to several different inputs (see section 6.8). Therefore, the assumptions implicit in economic theory may not always apply to agriculture, and it is important to examine the peculiar characteristics of agricultural inputs to see how they affect the operation of the industry and create problems for policy makers.

One of the main difficulties for the policy maker is that the problem is only rarely purely economic. Decisions designed to affect the ownership of agricultural land or the wages of farm labour may have implications for rural society or the rural landscape. And when they have to be taken for all 15 countries of the EU, geographically spread from Greece to Scandinavia and including the small farms of Portugal and the former collective farms of Germany, they become more complex still.

6.2 Some economic theory

Why are farm workers generally reckoned to be among the low-paid? Why do would-be entrants to farming often complain of the high price of land? What, in other words, determines the price of or reward for an input?

We can answer these questions by using the same ideas of demand and supply that we used to explain why the prices of farm products are what they are. But this time we are not looking at *product* prices but at *input* prices. So we are interested in the demand for the input, and what determines it, and the supply of the input, and what affects that.

The *demand* for an input is determined by:

(1) The quantity of output which will be produced by the input. Thus, we would expect that there would be more demand for fertile land than for infertile land, and for fit, well-trained workers than for sickly, inexperienced workers who will have lots of time off and break machines when they do manage to stagger to work.

(2) The price of the *product* which is produced by the input. If the price of the product is high, lots of firms will want to produce it, so there will be plenty of demand for the inputs required.

Thus, the demand for an input is often said to be a *derived* demand – it is derived from the demand for the product. If nobody wants the product, nobody will want the inputs used to produce it.

The *supply* of an input is largely determined by its price. The more firms are prepared to pay, the more land, labour and capital they will be able to acquire. It's not quite as simple as that because there are some other considerations: people may be willing to do pleasant or interesting jobs for lower wages than they would require to persuade them to do boring or dirty jobs; farmland cannot become building land, no matter what the price offered, without planning permission; and training programmes increase the quantity of labour able to do a job at any particular price level.

Fig. 6.1 shows that the price of an input (P_i) will therefore be determined by the point at which the demand (D_i) and supply (S_i) curves intersect.

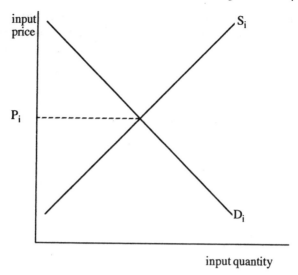

Fig. 6.1   Determination of the price of an input (P_i). D_i = demand for the input; S_i = supply of the input.

This theory implies that the rewards to inputs produced by the market have nothing to do with fairness or justice and everything to do with relative scarcity. Fig. 6.2 shows that when farm product prices increase the demand curve for agricultural inputs will shift to the right, from D_{i1} to D_{i2}. The reward to the input will rise (P_{i1} to P_{i2}) and so will

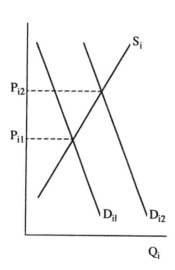
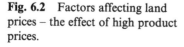

Fig. 6.2   Factors affecting land prices – the effect of high product prices.

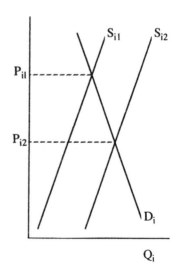

Fig. 6.3   Factors affecting land prices – easier planning permission.

the quantity employed. This explains the increase in land prices and rents when cereal prices are high. It can be seen as the result of competition for land on which to grow cereals and thus obtain the rewards from doing so.

Equally, Fig. 6.3 shows that if the difficulty of obtaining planning permission is reduced, the supply curve of potential building land will shift to the right (S_{i1} to S_{i2}) because the supply of land increases, and so its price is likely to fall as the quantity available on the market (Q_i) increases.

⇨ **Questions**_____

You can review your understanding of the theory with the following:

1   Why is there now little demand for wheelwrights?
2   What effect would an increase in pig prices have on the rent of farm land?
3   What effect would a 500% increase in grants for broadleaved forestry in lowland areas have on the price of land?

6.3   The structure of agriculture

The structure of agriculture is one of the main reasons for the problems of the CAP. Many, but not all, European farms are small. If they were all big, or all small, there would be fewer problems. But since they are such a mixture, the sort of policy which is appropriate for big farms causes problems for the small farms, and vice versa. Moreover, the mixture is not evenly spread: some countries, or regions within countries, have more big farms than average while others have a preponderance of small farms, so that aiming policy at one size group or another also produces nationalistic and regional arguments. The purpose of the next few paragraphs is to:

● Explain in more detail why size matters
● Explain how to measure and describe structure
● Describe the structure of EU agriculture
● Explain why the main structural trends are occurring
● Examine the implications of structural change.

From the policy maker's viewpoint, the important thing about big farms is that they can provide a living for a farm family at lower price levels than those required for equally efficient small farms. A bit of simple arithmetic will explain why.

	Small farm	**Big farm**
Total inputs (£)	10 000	100 000
Output (£) per £ of input	1.2	1.2
Total output (£)	12 000	120 000
Profit (£)	2 000	20 000

Of two equally efficient farms (in terms of output per £ of input), only the larger one provides a reasonable living. Thus, the perennial argument about the relationship between farm size and efficiency (i.e. output/input), interesting though it is, is perhaps less important for the policy maker or, as far as survivability is concerned, than the relationship between size and profitability (i.e. output minus input). The bigger the farm, the less efficiently it needs to be run, or, alternatively, the lower the price at which the farmer can make enough money to support a family.

The following diagram indicates the problem of organising policy to fulfil the needs of both small and large firms.

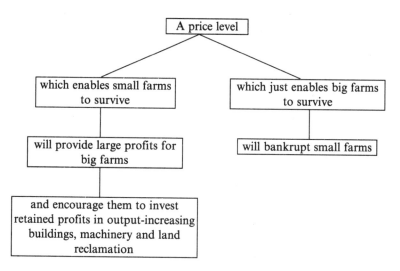

If the farms in an economy are all much the same size, the policy maker knows that setting a particular price level will have much the same effect on all of them. But with a mixture of farm sizes, high prices will keep everybody in business but run the risk of producing surplus output. Low prices will guard against surpluses but run the risk of putting small farmers out of business.

Thus, we can see that farm size needs to be measured in terms of size of business, and that we need to know whether the sizes of farm businesses in the EU are similar or vary from one country to another.

At first sight, it is easy to measure farm size, but the more you think about it the more difficult it becomes. The most obvious way is the one we use all the time – for example, 'This is a 100 hectare (250 acre) farm' – but you soon realise than 100 hectares of Grade 1 land in the Fens will support a much bigger business than 100 hectares of steeply sloping rough grass and heather on the side of a Scottish mountain. So what are the other options?

Ideally, we would measure the turnover of every farm business in the EU. The problem with that is that farmers would be aware of the implications of such data for their tax liability, and so might be unwilling to provide accurate information to anybody other than the tax authorities, who have an obligation of confidentiality.

All member states collect data on the acreage of crops and headage of livestock on each farm (in the UK it has been done by the June census since 1866). We could simply add up these figures for each farm, but there are no common units for acres of sugar beet and numbers of sheep, and in any case an acre of sugar beet is worth more than a single lamb.

In that case, we could multiply all crop acreages and livestock numbers by a standard labour requirement, assuming a rough equivalence of labour input and value. This was the approach used in the UK's *standard man day* system which was used for farm size statistics in the 1960s and 1970s.

Another option would be to multiply all crop acreages and livestock numbers by a standard gross margin. Clearly, this is not absolutely accurate, but it is probably good enough to distinguish the big farm businesses from the small ones and get around the major problems associated with the alternatives outlined above. It is the basis of the *European Size Unit* (ESU) system: one ESU is 1200 ECU of standard gross margin at 1987–9 values. Eight ESU is reckoned to be the minimum size for a full time holding.

Table 6.1 shows the structure of EU agriculture measured in hectares and number of farms, and the relatively large size of UK farms is apparent.

Table 6.2 shows the structure of EU agriculture in ESUs, and the picture becomes more complex, mainly because many Dutch and Danish farms are very intensive.

These tables show that most farms fall into the smaller size groups, whereas most land falls into the larger size groups, which suggests that the minority of bigger farms are responsible for most of the output from EU agriculture. In the UK in 1995, only 27% of all farms had more than 40 ESUs, but they accounted for 80% of the total ESUs in the country. In contrast, sub-8 ESU farms, although accounting for more than 40% of all farms, produced less than 3% of the total of

Table 6.1 Number of farms in each size group as a percentage of all farms (1989).

	1–5 ha	5–50 ha	50 ha and over
UK	13.5	53.1	33.3
Luxembourg	18.9	54.8	26.2
Denmark	1.7	81.0	17.2
France	18.2	63.6	18.1
Ireland	16.1	74.7	9.0
Netherlands	24.9	70.7	4.4
Belgium	27.7	66.5	5.8
West Germany	29.4	64.5	6.1
Italy	67.9	30.2	1.9
Greece	69.4	30.1	0.5
Spain	53.3	40.7	6.0
Portugal	72.5	44.0	6.8
USA	7.0	37.2	55.8
Australasia	13.1	26.4	60.5

Source: D. Grigg (1995) *An Introduction to Agricultural Geography*, 2nd edn., pp. 161, 163, Blackwell, Oxford.

Table 6.2 Percentage of farms in each ESU size class (1987).

ESU classes	0–8	8–40	40+
EU-12	72	22	5
UK	41	31	27
Denmark	7	65	29
Netherlands	4	51	44
West Germany	31	59	10
France	24	61	15
Italy	66	31	3
Spain	70	29	1

Source: M. Tracy (1993) *Food and Agriculture in a Market Economy*, p. 11, APS, La Hutte, Belgium.

ESUs. The same conclusion can be drawn from the figures in Table 6.3.

How can we explain this trend? Why is more and more of the output being produced by the bigger enterprises? Increasing efficiency of bigger units obviously has something to do with this increasing specialisation. Bigger combines, tractors or milking parlours can cope with more work than the smaller units they replace, and so need bigger enterprises if they are to be used most effectively. Bigger enterprises

Table 6.3   The proportion of output produced by the larger enterprises in the UK in 1993.

Farming enterprise	Proportion of total UK output (%)
Cereals produced on holdings ⩾ 50 hectares	68.6
Oilseed rape produced on holdings ⩾ 50 hectares	32.9
Sugar beet produced on holdings ⩾ 20 hectares	68.8
Potatoes produced on holdings ⩾ 20 hectares	49.0
Dairy cows in herds ⩾ 100	42.7
Beef cows in herds ⩾ 50	48.5
Breeding sheep in flocks ⩾ 500	49.7
Breeding pigs in herds ⩾ 100	79.3
Fattening pigs in herds ⩾ 1000	60.8
Broilers in flocks ⩾ 100 000	58.2
Laying fowls in flocks ⩾ 20 000	73.1

Source: MAFF (1994) *Agriculture in the UK 1993*, table 2.4, HMSO, London.

enable fixed costs to be spread over a bigger volume of output. Hence, the interest often shown by farmers in buying neighbouring fields when they come on to the market. They can usually be farmed with the labour and machinery already employed so the extra revenue produced is much more than the extra cost incurred.

But there is a limited to the extent to which greater efficiency can explain the increasing sizes of farms and enterprises. Hill and Ray argue that 'average efficiency improves with size up to the 2–3 man level, beyond which no further economies are evident, nor do dis-economies seem to appear'[1]. Nevertheless, as we have already seen, it will still be worth expanding output even though there are no efficiency gains to be had because output *minus* input – the total profit of the farm – will increase. At the same time, we should also remember that what is claimed to be the biggest family farming business in Britain, JSR Farms in East Yorkshire, a £15 million business employing 216 people in 1996, produces less than 0.2% of the UK agricultural industry's output and uses less than 0.05% of its labour force[2].

There is some controversy over whether or not these trends should be encouraged or resisted. The economic advantages of larger units are contrasted with the compensating advantages of smaller units, which are said to be vital to the maintenance of the rural economy, society and environment. Whether or not this is really the case is discussed in more detail in Chapter 8. For the present, suffice it to say that some aspects of agricultural policy do indeed maintain smaller farms, and others do not. What is certain is that the overall number of

Table 6.4   Total of farmers and farm workers in the EU ('000s).

	1960	1975	1990
Belgium	367	140	119
Denmark	—	177	147
West Germany	2216	1234	1081
Greece	—	—	889
Spain	—	—	1496
France	3426	1950	1394
Ireland	—	325	173
Italy	4007	2826	1913
Luxembourg	22	12	6
Netherlands	363	254	289
Portugal	—	—	840
UK	—	626	577
EU-6	10 402	6414	4802
EU-9	—	7542	5669
EU-12	—	—	8923

Source: Commission of the European Communities (1993) *Our Farming Future*, p. 16, The Commission, Brussels.

people working in agriculture in the EU is in decline, as Table 6.4 indicates.

⇨ **Questions**_____

Now answer the following to check your understanding of the structure of agriculture:

4   Why can it be misleading to measure farm size in hectares or acres?
5   What do we mean when we say that many Dutch or Danish farms are intensive?
6   Why is a good big farm likely to provide a better living than a good little farm?

6.4   **Substituting land, labour and capital**

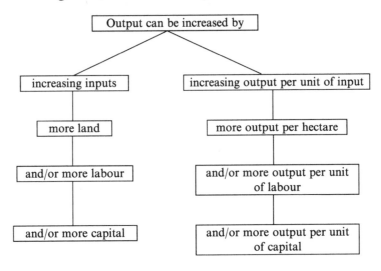

To some extent, land, labour and capital are substitutes for each other. We can increase the output from land by cultivating it more intensively, which means using more of the other inputs on it. In this case, the output per hectare of land rises, but the output per unit of the other inputs may fall. We can make labour more productive by giving it more machinery to work with: you can harvest more acres per day with a combine than with a scythe. On the other hand, if we introduce an innovation, such as a new variety, we may be able to increase the output per unit of land or labour without increasing the input of either. And if we use more of all inputs we shall also increase output. The diagram above sums it all up.

A change in the output per unit of input is a *productivity* change; thus, we can speak of land, labour and capital productivity, and of the productivity of all inputs together, which is *total factor productivity*. Over the last 50 years, land (see Table 6.5) and labour (see Table 6.6, which probably understates labour productivity in recent years owing to the effect of price changes) productivity have increased, but the evidence for increasing capital productivity is much more equivocal. This is because extra capital accounts in part for increasing land and labour productivity, and because it is much more difficult to measure capital productivity than land or labour productivity.

Estimates of overall efficiency (total output/total input) are not easy to make, because it is necessary to value both outputs and inputs in order to add them all together, and conflicts can arise over the correct price at which to value outputs. The obvious figure – the reigning

Table 6.5 Changes in the output of cereals (tonnes) per hectare of land in the UK and other EU countries.

Year	UK	France	Netherlands	Italy
1950–4	2.56	—	—	—
1970	4.01	3.38	3.76	2.69
1990	6.17	6.07	6.93	3.84

Sources: Commission of the European Communities (1993) *Our Farming Future*, p. 7, The Commission, Brussels; H.F. Marks and D.K. Britton (1989) *A Hundred Years of British Food and Farming: A Statistical Survey*, p. 164, Taylor and Francis, London.

Table 6.6 Output in constant (1986) prices per head of labour engaged in UK agriculture.

Year	Output (£'000s)
1950	9.8
1960	14.0
1970	20.0
1980	19.9
1987	19.3

Source: H.F. Marks and D.K. Britton (1989) *A Hundred Years of British Food and Farming: A Statistical Survey*, pp. 138, 150, Taylor and Francis, London.

market price in the EU – is normally higher than world price, which represents the cost of obtaining the product from an alternative source, and neither price takes account of the environmental effects of agricultural production.

6.5 Land as an input

The use of land as an agricultural input is complicated by several factors: it is not all of the same quality, there are other uses for the land (urban development, roads, leisure, nature conservation, forestry, water gathering), there are various legal advantages and problems in renting or owning it, and its price cannot always be explained by purely economic considerations.

The quantity of land in agriculture changes only slowly, but it is certainly not fixed. Some land moves out of agriculture into alternative uses. Paradoxically, it tends to be the better land which goes into urban development and the less good which goes into forestry and nature conservation. At the same time, rough grazing might be transferring into permanent grass or even tillage through reclamation. Table 6.7 summarises the changes in recent years.

Table 6.7   Land use in the UK ('000s hectares).

	1950	1970	1990
Total UK area	24 400	24 410	24 085
Total agricultural area	19 518	19 124	18 530
Total crops and grass	12 597	12 143	11 335
(of which arable)	(7 428)	(7 199)	(6 657)
Rough grazing	6 921	6 692	5 840

Sources: H.F. Marks and D.K. Britton (1989) *A Hundred Years of British Food and Farming: A Statistical Survey*, Taylor and Francis, London; MAFF (1994) *Agriculture in the UK, 1993*, Table 2.1, HMSO, London.

Some commentators have expressed disquiet over the loss of land from agriculture, but others have agreed that the increase in output per hectare more than makes up for the decrease in the total area. From another viewpoint (as is clear from Table 6.7), since nearly a million acres have been lost from the rough grazing area, some of which will have gone to forestry and some into more intensive agriculture, others have argued that it is the multiple use land which has been under greatest pressure. Rough grazing not only supports sheep and cattle but is also an important wildlife habitat and is used for all sorts of recreation, from walking to hunting and shooting.

The other major change in this century has been in the ownership and occupation of land. As Table 6.8 demonstrates, much more is now farmed by the people who own it than was the case in 1914.

Table 6.8   Tenure of farm land in Great Britain.

Year	Rented (%)	Owner-occupied (%)
1914	89	11
1950	62	38
1970	43	57
1986	39	61
1993	36	64

Sources: H.F. Marks and D.K. Britton (1989) *A Hundred Years of British Food and Farming: A Statistical Survey*, pp. 135–6, Taylor and Francis, London; MAFF (1994) *The Digest of Agricultural Census Statistics, 1993*, pp. 7–19, HMSO, London.

It must be remembered that the figures in Table 6.8 refer to the area of land. Not all farms are exclusively rented or owned; some are a mixture of the two and, in 1986, roughly half of all farms were owned, a quarter tenanted, and a further quarter part owned and part tenanted (i.e. in mixed tenure). In most countries of continental Europe, owner-occupation is as dominant, if not much more so, than it is in the UK. Whether or not this is a good thing is another question. 'The magic of property turns sand into gold' argued Arthur Young, the eighteenth century agricultural writer, but it might also be argued that owner occupiers have to find the money to acquire their farms, and so might be short of working capital, or even unable to enter farming at all. Then again, tenants do not benefit from the capital gains produced by rising land prices (see Table 6.9). It should also be remembered that changes in legal and economic circumstances can change the picture quite rapidly, as, for example, when the Agricultural Tenancies Act 1995 created Farm Business Tenancies (FBTs). The increasing pressure on small farm businesses is giving rise to a new type of tenure in which farming companies either rent land on a short-term basis (sometimes using FBTs) or farm it for a share of the profits.

Table 6.9   Land prices in England.

Year	Price (£/ha)	
	At current prices	In 1985 money values
1950*	153	1 739
1960*	249	1 886
1970	526	2 684
1980	3 470	4 908
1985	3 871	3 871
1990	4 683	3 513
1994	4 212	2 764

Sources: H.F. Marks and D.K. Britton (1989) *A Hundred Years of British Food and Farming: A Statistical Survey*, p. 137, Taylor and Francis, London; MAFF (1996) *Agriculture in the UK, 1995*, table 7.1, HMSO, London. Note that the current prices for the two years marked with an asterisk are for all farm land, whereas the other figures are for land with vacant possession.

These changes in the real value of land can be explained by using the theory outlined in section 6.2. If the physical output of land increases faster than the price of farm products decreases in real terms (as seems to have happened before 1980), then the derived demand curve for land will shift to the right (D_{land1} to D_{land2}) and its price will rise, as Fig. 6.4

demonstrates. If the real product price decreases faster than output increases (as seems to have happened after 1980), the land price will fall, as illustrated in Fig. 6.5.

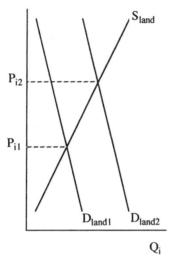

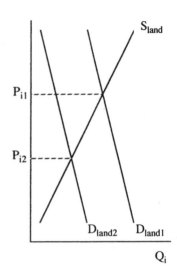

Fig. 6.4   Land prices before 1980.   **Fig. 6.5**   Land prices after 1980.

6.6   Labour as an input

Labour in agriculture means farmers and farm workers. Farm workers are among the most mobile of all the inputs employed in agriculture (which means they are most likely to leave the industry when times are hard), and farmers (in the UK, but not necessarily in continental Europe) are among the least mobile. Not surprisingly, therefore, the study of farm labour has mainly been concerned with numbers, and why they change, and the characteristics of farm labour. Clearly, these issues are interrelated. Moreover, they have implications for rural sociological and political questions. If numbers are declining, what will be the effect on village life and national agricultural politics?

The complexity of the farm labour force is clear from Table 6.10; it is not enough just to distinguish between farmers and farm workers. The decline in the number of whole-time hired male workers – the traditional farm workers – has not been matched by those in other categories. To some extent, the disappearing whole-time workers have been replaced by seasonal workers and spouses, whose numbers have increased. Within the virtually unchanged number of farmers, partners and directors of farming companies there is a decrease in the number of full time and an increase in the number of part-time farmers. The same

Table 6.10 Numbers of people engaged in agriculture (thousands).

Labour force	1975	1995
Workers		
Regular full time	222	103
Hired male	157	70
Hired female	15	10
Family male	37	20
Family female	13	3
Regular part-time	80	56
Hired male	22	19
Hired female	26	18
Family male	15	13
Family female	18	7
Seasonal or casual	73	84
Male	41	57
Female	32	27
Salaried managers	7	8
Total employed	382	251
Farmers, partners and directors	280	282
Full time	212	170
Part-time	68	112
Spouses of farmers, partners and directors (engaged in farm work)	75*	75
Total labour force	737	608

* Estimated.
Sources: A. Burrell, B. Hill and J. Medland (1990) *Agrifacts*, p. 40, Harvester Wheatsheaf, Hemel Hempstead. MAFF (1996) *Agriculture in the UK, 1995*, p. 15, HMSO, London.

pattern is not necessarily found in other member states of the EU, as Table 6.4 shows.

There are two explanations for the mobility of hired workers. One is that they are *pushed* out of the industry by farmers responding to falling real prices and so wishing to reduce costs (see Fig. 6.6); the other is that they are attracted to other industries by the higher rewards available there. This is the *pull* explanation (see Fig. 6.7). In each case, the number of employed workers decreases from Q_{i1} to Q_{i2}.

In Fig. 6.6 the demand curve for workers (D_w) is shifted to the left (D_{w1} to D_{w2}) by the decreasing real price of farm products, whereas in Fig. 6.7 the supply curve (S_w) is shifted to the left by the availability of better-paid jobs in other industries, so that at any given wage level there

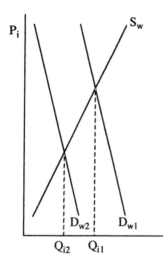

Fig. 6.6   Pushing workers out of agriculture.

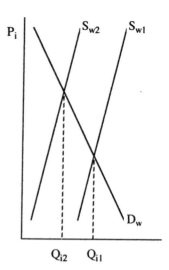

Fig. 6.7   Pulling workers out of agriculture.

are fewer people willing to apply for farm work. When general unemployment throughout the economy is high the push mechanism is most likely to operate, and when there are plenty of jobs available the pull mechanism provides the best explanation for the declining farm labour force.

6.7   Capital as an input

Capital in agriculture means buildings, machinery, breeding livestock and other forms of working capital. As the use of labour has declined, so the use of capital equipment has increased but, as Table 6.11 makes clear, most of this capital is still provided by the agricultural industry itself. It has over £60 billion worth of assets, but owes only about £8 billion to the banks and other sources of capital such as the Agricultural Mortgage Corporation (AMC). Nevertheless, this comforting ratio could be changed quite quickly by variations in the value of farm land, which constitutes the major part of the assets.

Since agricultural provides most of its own capital, investment in land, buildings and machinery is largely dependent upon the profitability of the industry. When profits are high, farmers go out and buy land and farm machinery; when they aren't, they don't. Some (e.g. Sir Richard Body) have argued that agriculture is over-capitalised, and that only the high-priced regime in which it operates, coupled with past encouragement in the form of grants for buildings and fixed equipment, have encouraged farmers to invest as much capital as they have. There certainly seems to be little doubt that some of the land and labour productivity gains have been produced by the investment of capital; whether they have been bought at the expense of capital efficiency is more difficult to determine.

Table 6.11 Estimated balance sheet for UK agriculture (£ million).

	1975	1994
Assets		
Fixed		
Land and buildings	12 350	43 400
Machinery	2 100	7 550
Breeding livestock	1 700	4 300
Current		
Trading livestock	1 250	3 500
Crops and stores	1 200	2 650
Debtors, cash, deposits	900	3 700
Total assets	19 500	65 050
Liabilities		
Long and medium term		
Bank loans	200	1 600
Other	700	1 800
Short term		
Bank overdrafts	750	2 700
Other	550	2 250
Total liabilities	2 200	8 350
Net worth	17 300	56 700

Sources: A. Burrell, B. Hill and J. Medland (1990) *Agrifacts*, p. 40, Harvester Wheatsheaf, Hemel Hempstead; MAFF (1996) *Agriculture in the UK, 1995*, p. 82, HMSO, London.

⇨ **Questions**_____

You can check on your understanding of land, labour and capital inputs with the following:

7 What might have happened in the 1980s to make the price of land fall from its peak value in real terms?

8 What would you expect to happen to the number of hired workers in agriculture if a sharp rise in world prices coincided with industrial depression in the UK?

9   If agriculture has so much capital, why do farmers worry about borrowing?

10 Why might farm machinery manufacturers lobby governments to provide farm income support?

6.8   Farm incomes: the rewards for inputs

Small farmers may use their own land, labour and capital to produce their output; large tenant farmers use the landlord's land, hire workers, perhaps borrow money from the bank to buy their capital equipment, and so produce their output by putting together other people's land, labour and capital. Consequently, calculating the return to the various inputs used in UK agriculture – their incomes – is not always a simple task. But each year, when the Ministry of Agriculture, Fisheries and Food announces the latest figure for farm income, there is much public interest, often accompanied by items in the news media and editorials in the trade press. So how is this important figure calculated? The following table takes you through the income estimate for 1995 (figures from MAFF (1996) *Agriculture in the UK, 1995*, pp. 66–69, HMSO, London).

Output (in £million)

of crops	6 949
+ livestock and livestock products	10 239
+ growth in breeding livestock	–64
+ government subsidies (e.g. set-aside payments)	328
+ change in value of stocks and work in progress	–14
= Gross output	17 438

Minus input costs (feeds, seeds, pesticides, fertilisers, fuel, vets, etc.)	8 042
Minus change in value of feed and fertiliser stocks	–35
= Gross product	9 361

Minus depreciation of buildings and machinery	1 855
= Net product	7 505

Minus net interest paid on loans (the cost of borrowed capital)	594
Minus net rent paid for tenanted land	151
= Labour income from agriculture	6 761

Minus the cost of hired labour	1 716
= Total income from farming	5 045
Minus the cost of family labour, and non-principal partners and directors	1 017
= Farming income (i.e. the reward for physical labour, management activity and the farmer's own capital)	4 028

This represents a return of a little over 6% on the £65 billion of assets involved in UK agriculture (see Table 6.11).

6.9 Summary

- Inputs are conventionally divided into land, labour, capital and management.
- The prices of inputs are determined by the demand for and supply of them.
- The size structure of agriculture has a big impact on the rural economy and society, and there are big differences in size structure between different European countries.
- Land prices and rents can be explained by economic considerations, but only up to a point.
- Labour is the input which is most likely to leave the agricultural industry.
- Land and labour productivity increases have been brought about by increased use of capital, and it is difficult to determine capital productivity.
- The overall income of the agricultural industry can be broken down to show the returns to the various inputs.

Notes

1. B. Hill and D. Ray (1987) *Economics for Agriculture*, p. 245, Macmillan, London.
2. D. Cousins (1996) 'Britain's Biggest and Still Growing', *Farmer's Weekly*, 17 May, p. 82.

Chapter 7
International Trade in Agricultural Products

7.1 Introduction

The EU is both a major importer and a major exporter of agricultural products, so you cannot understand how agricultural policy works without knowing something about international trade in those products. Moreover, international negotiations on trade problems have often been long, drawn-out and occasionally acrimonious, with agricultural trade problems affecting wider issues, as happened in the GATT (General Agreement on Tariffs and Trade) negotiations in the late 1980s and early 1990s. But even without these political problems it would still be important to know something about international trade, simply because the food processing and retailing firms which buy the produce of UK farms could also buy from producers abroad. Therefore, the purpose of this chapter is to explain:

- Why trade occurs
- Why the pattern of international trade in farm products is not always what you might expect it to be
- What the main agricultural trade problems are and why they occur
- How the GATT agreement fits in to the picture.

7.2 Why trade occurs

When people first began to farm, it seems, they probably produced what they ate, drank and wore without needing to trade much. So why bother with trade at all? In fact, prehistoric farmers probably *did* engage in trade for such items as salt and metals and, by the time of the Roman invasions, Britain was an exporter of grain and meat (which was one of the reasons for the invasions). Medieval Britain exported wool and imported spices and, in the seventeenth and eighteenth centuries, sugar, tea and coffee importation began to reach a significant scale. So it is clear that one reason for trade is to enable us to consume things that we cannot produce for ourselves, from fertilisers to pineapples and bananas. However, from the end of the nineteenth century Britain began to be a major importer of cereals, meat and dairy products, all of which we could quite obviously produce for ourselves. (And, in fact, Britain also produced pineapples in the nineteenth century, in heated greenhouses on the estates of rich people). There are two sets of arguments as to whether or not this is a good idea, one in favour of free trade and imports, and the other in favour of protection (prevention of free trade to protect domestic producers).

In favour of imports

What is the point of importing something which you can produce for yourself?

- The economists' answer to this question is *the theory of comparative advantage*, which is explained in the appendix to this chapter. If a country has a comparative advantage in the production of a product it will maximise its income by producing that product and exporting it, in return for imports of products in which it does not have a comparative advantage. *This concept is fundamental to the understanding of international trade, so it's a good idea to read the appendix before you go much further.*
- Even where no obvious comparative advantage exists, specialisation may lower costs through economies of scale, so producing a comparative advantage.
- Conversely, protection (i.e. prevention of free trade) may protect inefficient producers from competition from foreign suppliers, which may suit them but results in domestic consumers having to pay higher prices than they would have to pay if they could buy the cheaper, imported goods.
- Moreover, protection from one country tends to provoke a similar response from other countries – so called 'beggar my neighbour' policies. Clearly, if UK consumers buy French wine then French wine growers will make profits which they may spend on lamb produced in the UK, but if wine sales fall there will be no profits to spend on lamb.

Generally speaking, therefore, economists argue that free trade makes people better off than they would be under protectionist systems.

In favour of protection

Some of the arguments in favour of protection are bad for most people but good for some people:

- Such as the argument about protecting UK workers from cheap foreign labour. If goods can be produced more cheaply abroad but are kept out of the country by import controls, the UK workers who produce competing but higher-priced products keep their jobs and so benefit, but at the expense of UK consumers who have to pay more than they would if they were allowed access to cheap imports.
- Or the argument about protecting declining industries. Strictly speaking, according to the way that markets are supposed to operate, inputs in declining industries should move out into expanding industries. But inputs are not always as mobile as that. It may be quite easy to move a tractor from the agricultural to the construction industry, it simply requires a farmer to sell it to a builder. But it may not be quite so easy to move a farm worker into the expanding financial services sector, because the skills required in one industry may not be useful in the other. Therefore, some protection, to allow declining industries to decline more slowly and so reduce the human costs of rapid adjustment may be justified, although again the consumers will be denied access to cheaper imports.

Some of the arguments are good for most people:

- Such as the argument for protecting infant industries, which are those which are just beginning, and have not yet achieved sufficient economies of scale to compete internationally. The problem, of course, is to decide when the infant should have grown up.
- Or the argument about the prevention of dumping (which is selling goods at less than their cost of production by using export subsidies).
- Or the argument for preventing the importation of harmful products, such as diseased plants or animals. The problem with this is that it can sometimes also be used to keep out perfectly healthy plants and animals to protect domestic producers.

And some of them are political, rather than economic, arguments:

- For example, the desire to maintain self-sufficiency in food and weapons in case of war.
- The feeling that ways of life based on traditional industries should be preserved.

Trade barriers

There are two sorts of trade barriers: tariff barriers and non-tariff barriers (NTBs). Tariff barriers include import taxes and export subsidies (measures which affect the prices of imports and exports). NTBs include environmental and safety regulations (for example, some countries do not allow sales of lawn mowers which make too much noise) and veterinary and phytosanitary (plant health) regulations.

In the depression years between World Wars I and II, many countries introduced trade barriers to protect their domestic industries. Since World War II, many countries have tried to reduce their trade barriers as the arguments for free trade have become more widely accepted (or, as some have argued, since powerful countries have realised that free trade is in their interests and have managed to impose their views on other countries). In addition, many new international organisations were set up, one of which lead to 23 countries signing the General Agreement on Tariffs and Trade (GATT) in 1947.

Under the terms of the GATT, member states have met periodically to negotiate reductions in trade barriers. These meetings are known as *negotiating rounds*, of which there have been eight. The latest was the Uruguay Round, so called because it began in 1986 at Punta del Este in Uruguay. Over 100 countries, representing 90% of world trade, were involved in the talks. The Uruguay Round was much more comprehensive than its predecessors, which had concentrated mainly on reducing trade barriers for manufactured goods (with some success, as average tariff rates had fallen from 40% in 1947 to about 5% by the end of the 1980s). At the talks in Punta del Este, it was agreed that the Round should also cover services, intellectual property (such as copyright) and agriculture. The negotiations went on until December 1993 and resulted in the creation of a new World Trade Organisation (WTO) and further agreements on trade. Agreements on agriculture were also made and the effect of these are mentioned later in section 7.5. The implementation of the GATT agreement is the responsibility of the Committee for Agriculture of the WTO.

Having read this section, you should now be able to see that economic theory suggests some clear conclusions about agricultural trade:

(1) Countries are likely to produce products in which they have a comparative advantage.
(2) Low income countries of The South are unlikely to have a comparative advantage in the production of manufactured goods, so their comparative advantage should be in agricultural production.

(3) Therefore, you would expect low income countries to export agricultural products and import industrial goods, while industrialised countries would be expected to import agricultural products and export industrial goods.

In fact, the real world is more complicated than that, and some of the major exporters of agricultural products are industrialised (often also referred to as 'developed') countries such as the United States and the member states of the European Union. We shall examine this apparent puzzle in more detail in the next section.

If you have not yet done so, you should now read the appendix on the theory of comparative advantage.

⇨ **Questions**_____

1 From what you already know of the CAP, would you say that it encourages free trade or that it is protectionist, as far as agricultural products are concerned?
2 Which of the arguments outlined in this section have been used to justify protectionism for EU agriculture?
3 If Germany attempts to exclude UK beef exports on the grounds that there is a danger of them being affected by BSE (bovine spongiform encephalopathy), is this a tariff barrier or a non-tariff barrier?

7.3   The pattern of agricultural trade

Table 7.1 shows some of the more important commodities entering into international trade, arranged in order of the total value of world exports.

Several interesting points emerge from Table 7.1:

- Trade in foods generally represents only a small proportion, usually less than 20% and sometimes much less, of world production. The only significant exceptions to this are sugar and oilseeds.
- In contrast, tropical beverages – coffee, cocoa and tea – are exported to a much greater extent, as are industrial raw materials such as cotton and rubber.
- The major exporters of many of the most traded products are industrialised countries, which is *not* what the theory of comparative advantage would lead you to expect.

- The major importers of many of the most traded products are industrialised countries, which *is* what the theory of comparative advantage would lead you to expect (although wheat and coarse grains are glaring exceptions to this rule).

Table 7.1　World agricultural trade in 1992 and 1993.

Product	Value of world exports ($'000m)	Total exports (% of world production)	Percentage of world exports *from* industrialised countries	Percentage of world imports *by* industrialised countries
Meat	19.83	5.5	74.3	62.3
Wheat	19.02	16.4	88.8	27.6
Coarse grains	14.66	10.4	75.2	46.2
Sugar	10.60	28.2	33.3	55.3
Wine	9.00	17.2	95.6	93.7
Oilseeds	8.71	33.8	45.9	51.4
Cotton	8.24	28.7	58.8	37.4
Coffee	7.24	80.4	4.3	91.2
Citrus fruit	6.25	10.0	73.0	90.9
Tobacco	5.86	19.8	41.1	40.2
Rice	4.27	2.8	24.1	21.4
Cocoa	3.76	77.4	0.0	58.7
Rubber	3.63	75.5	1.6	69.3
Bananas	2.93	no data	7.2	90.9
Tea	1.82	42.6	0.0	49.5
Cassava	1.06	5.9	0.0	76.9

Source: Calculated from data for 1992 and 1993 in FAO (1994) *Commodity Review and Outlook, 1993–4*, FAO, Rome.

⇨　**Question** _____

Try using Table 7.1 to fill in the boxes in the table on page 88:

4　If less than 50% of the exports of a commodity are from industrialised countries it will go in the right hand column, and if more than 50% of the imports are by industrialised countries it will go in the upper row. Thus, bananas appears in the top row in the right hand column and coarse grains in the bottom row in the left hand column

　　When you have put all the commodities from Table 7.1 in the appropriate box, calculate the total value of trade in each box.

The major trade flows

Majority of imports to	Majority of exports from	
	Industrialised countries	Low income countries
Industrialised countries		Bananas
Low income countries	Coarse grains	

Within the overall world picture, some countries are particularly important as exporters and importers. The USA, Canada, Argentina and Australia are traditional grain exporters, New Zealand has exported lamb and dairy products since the end of the nineteenth century (in fact, many of these traditional trade patterns have only been traditional since the 1880s), and some of the EU countries, especially the UK, were traditional importers. But there have been changes in recent years. Since the 1960s, there has been a major change in the pattern of trade in wheat and coarse grains (which, as Table 7.1 shows, are two of the major traded commodities). The EU countries have joined the list of significant exporters, and overall exports have increased in volume terms, as Table 7.2 shows.

In the same period, developing world countries have increased their imports of wheat and coarse grains. Nevertheless, the overall conclusion that world agricultural trade is dominated by the industrialised countries and that those countries account for a greater proportion of exports than comparative advantage theory would predict, remains valid.

One of the reasons why this is so is that some of the developed countries, and especially the EU countries, protect their agricultural industries. The extent to which they do so can be measured by the *Producer Subsidy Equivalent* (PSE), which is the annual monetary transfer to agricultural producers from domestic consumers and tax-

Table 7.2   Major grain exporters (percentage of total world exports).

	USA	Canada	EU	Total exports (million tonnes)
Wheat				
1960–64	42.6	24.2	0.0	45.9
1985–89	36.2	20.5	16.9	91.8
Coarse grains				
1960–64	52.5	3.5	0.0	29.7
1985–89	61.6	6.0	4.6	83.0

Source: Selected figures from D. Blandford, C.A. Carter and R. Piggott (eds) (1993) *North–South Grain Markets and Trade Policies*, p. 18, Westview Press, Oxford.

payers resulting from agricultural policy (see Table 7.3). It is calculated using the formula:

$$\% \text{ PSE} = 100 \text{ (Total PSE)}/Q(P_D) + D - L$$

$$\text{Total PSE} = Q(P_D - P_W) + D - L + B$$

where: Q is level of production;
   P_D is domestic producer price;
   P_W is reference price (or world price);
   D is direct payments to producers;
   L is producer levies;
   B is other budget payments to producers.

Table 7.3   Producer Subsidy Equivalents (%) for various areas/countries.

	1979–81	1992
EU-10/12	42.8	47
Nordic*	56.1	—
Mediterranean**	26.1	—
Japan	59.4	71
USA	16.0	28
Canada	23.9	44
Australia	4.7	12
New Zealand	15.5	3

Sources: OECD (1987) *National Policies and Agricultural Trade*, p. 117, OECD, Paris; OECD (1993) *Agricultural Policies, Markets and Trade*, p. 108, OECD, Paris.
* Finland, Iceland, Norway, Sweden, Switzerland.
** Portugal, Spain, Turkey.

A similar concept, called the *Aggregate Measure of Support* (AMS) was used in the Uruguay Round of GATT negotiations. The AMS differs from the PSE in excluding policy measures which do not affect producer prices or production levels. In other words, it is concerned with policies which are likely to distort the pattern of trade which would be produced by the free market[1].

As Table 7.3 demonstrates, traditional exporting countries generally have lower levels of support than traditional importers. The established EU member states and the new entrants, with the exception of the Mediterranean countries, generally have had high support levels. As a result, EU domestic production has increased, replacing imports and spilling over into the export market, as can be seen in Table 7.2 and, for the UK, in Table 7.4.

Table 7.4   UK self-sufficiency in food.

	Home production as % of total food consumption	Home production as % of indigenous food consumed
1964–5	50.9	64.1
1974	46.3	57.9
1984	62.9	82.6
1995	58.8	76.4

Sources: H.F. Marks and D.K. Britton (1989) *A Hundred Years of British Food and Farming: A Statistical Survey*, p. 121, Taylor and Francis, London; MAFF (1996) *Agriculture in the UK, 1995*, p. 5, HMSO, London.

The UK is now significantly below 100% self-sufficiency in only a few of the major products, specifically, sugar, fruit and vegetables, bacon and ham, and some dairy products (see Table 7.5). It has changed from being a major importer to being much less significant as an importer and not unimportant as an exporter, particularly to other EU member states.

7.4   The problems of international trade

In this section we shall be thinking about the *world market* in agricultural products. In fact, it is quite difficult sometimes to see exactly what is meant by a world market. International trade takes place between firms in different countries, and there is no one location or organisation in which trade takes place. Thus, the world market is, to some degree, an economist's abstraction. Nevertheless, for most agri-

Table 7.5 UK self-sufficiency in major agricultural products, 1995.

Product	Self-sufficiency (%)
Wheat	119
Barley	131
Oats	113
Oilseed rape	76
Linseed	63
Sugar beet	67
Potatoes	89
Tomatoes	29
Apples	35
Beef	107
Mutton and lamb	111
Pork	102
Bacon and ham	46
Poultrymeat	95
Butter	62
Cheese	73
Skimmed milk powder	177
Eggs	97
Wool	65

Source: MAFF (1996) *Agriculture in the UK, 1995*, HMSO, London.

cultural products, buyers have lots of potential suppliers in different countries and there are plenty of potential importers, so Spanish buyers of maize from the USA would expect to pay about the same price per ton as Japanese buyers at the American port. (This would be the f.o.b. (free on board) price. The grain then has to be transported to the importing country, and shipping and insurance costs will be incurred, so if these are included the price will be quoted on a c.i.f. (cost including insurance and freight) basis.) Thus, we would expect world prices to reflect the balance of world demand and supply, although it is important to remember that, for some products, individual countries might have such a preponderance of buyers or sellers that they can have a disproportionate effect on world market prices.

The recent disturbances to traditional trade patterns have produced some controversies in international trade negotiations, but international agricultural trade problems are not new and are as much a result of the underlying structure of production and trade as of any recent developments. It might in fact be argued that trade problems fall into two types:

(1) Those which would occur in the absence of government inter-
     vention.
(2) Those which are a result of government intervention.

Type 1 problems: those which occur *in the absence of government intervention*

They arise because, as we have seen in Table 7.1, only a small proportion of world production sells on the world market. Exporting
countries will normally wish to satisfy their domestic demand before
they export to other countries. Consequently, in years of shortage, they
will export less than normal and, in surplus years, they will export more
than normal. Now suppose that world production fluctuates by a small
amount – say 5% – each year. In a high output year, this implies that *an
additional* 5% of world production will be sold on the world market.
But for a product such as coarse grains, only about 10% of world
production is sold on the world market anyway, so the additional sales
represent a *50%* increase in exports. Similarly, in a shortage year, 5%
of world production will be withheld from the world market, which
means that world market supplies have decreased from 10% to 5% of
world production – *they have halved*. In short, there are big *volume*
fluctuations in world trade.

However, importers will probably want roughly the same amount
each year, and their price elasticity of demand is likely to be low (if you
can't remember why, see Chapter 3 on the demand for agricultural
products). So if we draw an inelastic demand curve, and show a big
fluctuation in the quantity on the market (Q_1 to Q_2) we can see what
will happen to price (see Fig. 7.1).

As Fig. 7.1 shows, big volume fluctuations give rise to even bigger
price fluctuations (P_1 to P_2). This creates a problem for importing
countries which probably want a steady level of supplies at predictable
prices, so they try to achieve this by entering into trade agreements or
by producing more of their supplies at home. This means that an even
smaller proportion of production sells on the free world market, which
takes you back to where you started and makes the world trade
problem more difficult again.

Type 2 problems: those which *result from government intervention*

If governments use protectionist measures (i.e. imposing trade barriers
to keep out imports, which is what the EU does) to support their
agricultural industries, the results are that:

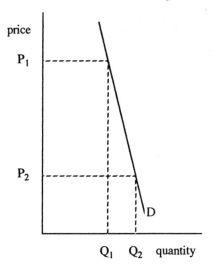

Fig. 7.1　Price fluctuations on international markets.

- Even less farm output goes on to the free world market, so it becomes more unstable.
- Consumers in protectionist countries may no longer buy from the cheapest supplier. They either buy from domestic producers or from producers in other countries in the trading bloc of which they are a member (such as the EU). See Table 7.4.
- Farmers in the protectionist countries sell their products for more than they would get on the world market and so produce more than they would produce at world prices. One of two things follows from this:
  - (a) either the additional output will replace imports, in which case the countries which previously supplied the imports will feel aggrieved because they have lost access to their traditional markets; or
  - (b) domestic production will increase still more, to the point where it exceeds domestic demand, in which case it will have to be sold on to the world market. Since it has been produced at higher than world market prices this will require an export subsidy, so traditional exporters will see this as unfair competition or dumping and domestic taxpayers will have to pay for the export subsidies. In some cases, surpluses are used as food aid, in which case the cost to domestic taxpayers is greater (and there is also an impact on producers in the aided country but that is too complex an argument to start at this point).

Agricultural economists have argued over whether or not it is possible to support farm incomes without having some impact on international trade. Clearly, the EU's support system relies on price support by protection and it therefore has an immediate and obvious effect, but even systems which are intended to leave domestic markets open to exports will have some effect, depending on the extent to which they encourage farmers to produce more than they would in the absence of support. Thus, the UK's deficiency payments system was designed to leave the UK market open to imports. However, since UK self sufficiency increased from about 30% to 60% between 1939 and 1970 it must have had some import-replacing effect. It is often argued that direct income payments to farmers would be 'production neutral' but, to the extent that they can be seen as reducing fixed costs, they may keep producers in business who would otherwise leave the industry.

⇨ **Question** _____

Now see if you can sort out this point:

5   By examining Table 7.1, can you identify any commodities which you would *not* expect to suffer from what we have called type 1 problems?

7.5   Solutions for agricultural trade problems

There have been many attempts to solve the problems of international trade, none of which has met with spectacular success. The most recent attempt, culminating in agreement in 1993, was the Uruguay Round of GATT negotiations. It was based on the idea that trade problems would be ameliorated, if not completely solved, by reducing the level of protection in world agricultural markets. In other words, importers agreed to reduce import barriers while exporters agreed to reduce export subsidies; in the terms of the discussion above, it might be seen as an assault on type 2 problems, those resulting from government intervention. The agreement affected three areas of the CAP:

(1)   *Internal support*, in which a 20% cut in trade distorting measures (i.e. those included in the AMS) over a six-year period was required. In practice, this required no changes in the CAP as it now exists since this level of reduction was incorporated into the 1992 CAP reforms.

(2)   *Import restrictions*, in which all existing variable levies, minimum import prices, voluntary restraint agreements and other import

restricting measures were required to be converted into tariffs, which would then be reduced by 36% over 6 years. In practice, the effect of this too was less than the changes agreed in the 1992 CAP reforms. There were also provisions for increasing access for imports for up to 5% of internal consumption, and this may have a greater effect on internal EU markets.

(3) *Exports*, for which a 36% reduction over 6 years in expenditure on export subsidies was required, together with a 21% cut in the volume of subsidised exports. Many commentators have argued that it is this last measure which will have the greatest effect on the prices received by EU farmers.

There was also an agreement on sanitary and phytosanitary measures, so that member states could take measures to protect human, animal and plant health, as long as these were based on scientific principles and not designed to create new barriers to trade.

As a result of these changes it was expected that world market prices for agricultural products would rise as the level of subsidised exports fell. By early 1996, world prices had risen but more as a result of decreases in supply and increases in demand in the faster-growing Asian economies than as a direct effect of the Uruguay Round. Before the agreement was concluded, farm lobbies were arguing that the proposals as they then existed would reduce farm incomes, but that has yet to be demonstrated in practice. Others went further: T. Clunies-Ross and N. Hildyard[2] argued that small farmers would be driven out of business while control of markets and inputs would be concentrated still further into the hands of multinational companies. Clearly, much depends on the evolution of world prices. It also appears that several countries see the Uruguay Round agreement as merely the beginning of a longer process of multilateral negotiation, still aimed at reducing the extent and impact of government intervention in world agricultural trade.

7.6 Summary

- The theory of comparative advantage suggests that free trade makes everybody better off but sometimes there are good reasons for protectionism.
- Comparative advantage suggests that third world countries should be the main agricultural exporters but, in fact, world agricultural trade is dominated by the industrialised countries.
- In recent years, the UK and the EU have increased their levels of

self-sufficiency, which has created some international political problems.

- There are two kinds of reasons for agricultural trade problems:
  (a) Problems arising in the absence of government intervention, because only a small proportion of world output is traded on the world market.
  (b) Those resulting from protectionist agricultural policies.
- The GATT agreement attempted to deal with the second kind of problem, and will have some impact on EU farmers because they have to reduce the quantity of farm products they sell with the aid of export subsidies.

Notes

1. OECD (1995) *The Uruguay Round: A Preliminary Evaluation of the Agreement on Agriculture in the OECD Countries*, p. 36, OECD, Paris.
2. T. Clunies-Ross and N. Hildyard (1992) *The Politics of Industrial Agriculture*, p. 106, Earthscan, London.

Appendix: the theory of comparative advantage

This theory was first expounded by the English economist David Ricardo (1772–1823). To explain how it works, let us take a completely hypothetical example using two mythical countries, East Angula and Souwestia. Table A7.1 shows what *one hectare* of land (and the necessary associated labour and capital) can produce in each of the two countries.

Table A7.1 Hypothetical production levels in two imaginary countries.

	East Angula	Souwestia
Wheat	8 tonnes	6 tonnes
Milk	12 000 litres	10 000 litres

From Table A7.1 we can see that:

(1) East Anglia produces more of both products; thus, it is said to have an *absolute advantage* in the production of both, and at first sight it might appear to be in Souwestia's interest to trade with East Angula, but not vice versa.
(2) A hectare of land in East Angula can produce *either* 8 tonnes of wheat *or* 12 000 litres of milk, but obviously not both.
(3) Similarly, a hectare of land in Souwestia currently producing

10 000 litres of milk might be diverted to wheat production, but only at the cost of that 10 000 litres of milk. So if Souwestia wants to consume wheat, it can either give up milk and use its own land to produce its own wheat, or it can buy the wheat from East Angula, paying for it with milk produced in Souwestia.

Presumably, the best policy is the one which allows Souwestia to obtain wheat at the lowest cost. We can measure this in *opportunity cost* terms: in other words, in order to obtain wheat, we have to divert a hectare from milk production to wheat production. How much milk do we have to give up to obtain a tonne of wheat?

In Souwestia:

to produce 6 tonnes of wheat we have to sacrifice 10 000 litres of milk, so to produce 1 tonne of wheat we have to give up (10 000/6) litres of milk

We might put this in albegraic terms as:

$6W = 10M$, so
$W = 10/6\,M$, where W = 1 tonne of wheat and M = 1000 litres of milk

We can go through this process for both products and both countries. The results of doing so are shown in Table A7.2.

Table A7.2 Opportunity costs of wheat and milk production.

	East Angula	Souwestia
Wheat	12/8M	10/6M
Milk	8/12W	6/10W

Now, which country has the lowest cost of wheat production? In fact, it's not easy to see immediately whether 12/8 is more or less than 10/6, so the table will be easier to use if we convert the figures for each commodity to the same denominator, which is what has been done in Table A7.3.

Table A7.3 Opportunity costs of wheat and milk production.

	East Angula	Souwestia
Wheat	36/24M	40/24M
Milk	40/60W	36/60W

From Table A7.3 we can see:

(1) That the lowest opportunity costs of wheat production are in East Angula (36/24 is a smaller number than 40/24). In other words, if Souwestia produces its own wheat it has to give up 40/24M (= 1666.7 litres of milk) for each tonne obtained, but if it imports from East Angula it only has to give up 36/24M (= 1500 litres of milk).

(2) Souwestia gets more wheat per litre of milk by importing from East Angula than by producing at home, and this is the advantage of international trade.

(3) This explanation of the advantages of international trade is known as *the theory of comparative advantage*, and East Angula is said to have a comparative advantage in wheat production. Table A7.3 also demonstrates that Souwestia has a comparative advantage in milk production.

Comparative advantages arise because countries have different inputs available. They differ in land fertility, climate, raw materials, population density, skills available in the labour force and capital equipment. Countries with warm, moist summers will find it easier to grow grass than cereals, and the reverse will apply to countries with hot, dry summers.

Thus, the whole point about the theory of comparative advantage is that it demonstrates that *both* partners in an international trading relationship will benefit, and that:

- The greater the opportunity cost differences the greater the gains from trade.
- The lower the transport costs the greater the gains (the theory ignores transport costs, or implicitly includes them in the opportunity costs). This is why there is more international trade in wool, which is high in value in relation to its bulk, than in potatoes, which are cheap and bulky and so relatively expensive to transport.

▷ **Questions** _____

Now test your understanding of comparative advantage by answering the following:

A1 Can you explain, from Table A7.3, why Souwestia has a comparative advantage in milk production?

A2 Here is an example from the beginning of history. A tract of land can either be quarried to produce menhirs, or it can be left in forest to produe wild boar but, obviously, not both. The production levels in two countries are shown in the following table:

	Gaul	Britannia
Menhirs	6	4
Wild boar	4	2

(a) Which country has an *absolute* advantage in the production of both products?

(b) Which country has a comparative advantage in menhir production?

(c) Which country has a comparative advantage in wild boar production?

(d) Will it be worthwhile for Britannia to produce its own wild boar?

Chapter 8
The Case for Supporting Farm Incomes

8.1 Introduction

As we have seen in previous chapters, agriculture faces a relatively static market for its products; at the same time its output increases as new technology is constantly introduced. Therefore, in a free market in which there is no government intervention, prices for farm products are likely to fall, so the income of the agricultural industry as a whole is likely to fall and so the incomes of individual farmers will fall as long as the number of farmers remains unchanged. In addition, as we have also seen, farmers, in the UK at least, do indeed tend to remain in agriculture unless placed under very severe economic pressure to withdraw.

In addition, since the late nineteenth century, farmers in western Europe have faced competition from producers in North and South America, Australia and New Zealand, who, by virtue of climatic and farm size structure differences, can profitably produce cereals, meat and dairy products, transport them to the European market and sell them at a price which is less than many European farmers require if they are to make a profit. In a free market with no government intervention, these farmers would, in the long run, either have to accept lower incomes or be forced out of agriculture. Their output would be replaced by imports from the lower-cost producers in other countries.

Faced with these predictions, or, in some cases, actual developments, governments have to choose whether or not to support farm incomes. *This chapter is concerned with the arguments which have been put forward to persuade them to offer or withold support*. It is *not* concerned with the merits or demerits of various *methods* of support. It is presumably preferable to decide, first, whether or not agriculture should be supported at all, and then to decide upon the best way to do it afterwards. The only problem with this is that some of the predictions about what would happen in the presence or absence of support depend upon the support method chosen. There is no way round this problem; we shall just have to take care to recognise when it might happen.

Numerous arguments for and against farm income support have been advanced over the years. Most of them fall into one of four categories:

(1) *Trade* arguments.
(2) The argument about whether or not farmers have an *income problem*.
(3) The argument about the effect of a declining agriculture on the *rural economy and society*.
(4) The argument about the effect of declining farm incomes and of farm income support on *the environment*.

We shall examine each of these in turn in the following pages.

(It is important to realise that this chapter is about farm income *support*, which is controversial, and not about the case for *stabilising* farm prices, which, while not accepted by all agricultural economists, is much less controversial.)

8.2   The trade arguments

Most of the trade arguments are about how *self-sufficient* a country should be. One side will argue that we should be more self-sufficient, and the other that we should rely more on imports for our supply of agricultural products. If there are no barriers to imports, consumers will buy imported goods in preference to home-produced alternatives if they are cheaper or more to the consumers' tastes. Thus, market forces will determine the proportion of imported goods sold in any country's markets. But, if you think about it, this gives us the possibility of various *definitions of agricultural self-sufficiency*, of increasing complexity:

*   The simplest is 'home production divided by total consumption', but the problem with this is that some of the food we consume consists of products like bananas, coffee and mangoes, which are tropical products and so are never likely to be produced in the UK.
*   We can overcome this problem by only including the products which we do produce on a significant scale in the UK. These are called *indigenous* products, so a further definition would be 'home production of indigenous products divided by total consumption of indigenous products'. The problem with this definition is that in the real world there are barriers to imports which increase self-sufficiency and export subsidies which reduce it.
*   In theory at least, we can overcome this problem by defining self-

sufficiency at a particular price level. Thus, the figures reported in Table 7.5 are at CAP price levels. At world price levels, the UK would produce less and more would be imported. But there is still a problem: if we found, for example, that at world price levels the UK was 50% self-sufficient, we might discover that this level of home production could only be supported by using imported fuel, fertilisers, and feedingstuffs.

- Therefore, some people would argue that real self-sufficiency should be defined as 'the proportion of total supply that a country can produce using only indigenous inputs, and operating at world prices'.

For the UK, figures for the first two definitions of self-sufficiency are given in Table 7.4, with figures for individual commodities appearing in Table 7.5. There are no published figures for self-sufficiency according to the third and fourth definitions. If all this seems a bit academic, read the next few paragraphs and see if its relevance becomes clearer.

To return to the story: we have seen that one of the arguments for supporting farm incomes is that farmers will be kept in business and producing farm products, whereas in the absence of support they would produce less and perhaps even go out of business altogether. Clearly, this would affect the level of national self-sufficiency in food. Now, the crucial question is *'does this matter?'* So what if self-sufficiency falls? So what if we are totally self-sufficient? As you might expect, there are some arguments in favour of greater self-sufficiency, and some against it.

The arguments *in favour* of greater self-sufficiency

Food security

We should look pretty silly if we couldn't feed ourselves in an emergency. This is a more complex issue than it appears at first sight, because we then have to define an emergency and predict our reactions to it. In an emergency, for example, we might be prepared to accept less meat in our national diet, as we did during World War II and, since you can feed more people from a hectare devoted to crops than from a hectare grazed by livestock, self-sufficiency would increase. What sort of emergency are we talking about? We have just mentioned a war, but wouldn't future wars, on a large enough scale to disrupt world markets, rapidly go nuclear? Modern farming methods rely on electricity, diesel oil, fertilisers and pesticides which are likely to be in short supply during and after a nuclear war.

Less apocalyptically, it is possible to imagine a world food shortage brought about by rapid population growth or crop or animal diseases in which case a country's ability to feed itself using its own inputs would become increasingly important.

In the everyday world, however, it might be suggested that regular trading relationships are a form of food security. If importing countries are worried about where their supplies are coming from, exporters are often just as worried about where their exports are going to. Access to markets for traditional exporting countries was a big issue in the Uruguay Round of GATT negotiations, and Sir Richard Body has gone so far as to suggest that the Falklands War of 1982 would not have happened if Britain had still been importing Argentinian beef, because Argentinian beef producers would have had too much to lose and so would have restrained their government[1].

Ethics

Some people think that if we have the capacity to produce food we should do so in order to feed the starving millions in the rest of the world. This is partly a matter of value judgement upon which everybody has to make their own decision, but there are other issues involved too. The main problem for the starving millions is not that there is no food on the world market for them to buy, but that they have no money with which to pay for it. People then have to decide whether paying farmers in industrialised countries to produce food for free distribution to starving people is the most effective way of solving the problems of the starving. In the long run, the money required might be better employed in enabling them to produce their own food.

Employment effects

The agricultural input industries and the food processing industries depend on agriculture and employ a lot of people. Thus, if agriculture is run down the effects are felt in other industries, and if agriculture expands there are jobs for people in other industries too. It's not quite as simple as that because the input industries – fertilisers, pesticides, agricultural machinery, for example – could still export their products in the absence of a home market for them, although whether they would sell as much is another question. Similarly, the food industries could still operate to process imported raw materials, although presumably foreign suppliers would try to add as much value as they could before exporting their products.

It is not difficult to explain why the employment effects of the agricultural industry exist; quantifying those effects is a more complex and time-consuming exercise. Some things are straightforward: for example, Tables 1.1 and 1.2 show the proportion of agricultural jobs in the economies of the UK and other EU member states. Recent data on agriculture-related employment in other industries is more difficult to obtain and the most easily available source (see Table 8.1) only has figures for 1981. However, it is still worth examining Errington's data. He estimated that the proportion of output sold directly to UK agriculture varied from 79.5% of the output of the animal feedingstuffs industry, and 60.1% of fertiliser output, to only 0.9% of motor vehicle output and 0.5% of the output of the banking, finance and insurance industries. He also estimated agriculture-related employment in the UK food processing industry, and so produced the figures in Table 8.1, which suggest that there are roughly the same number of agriculture-related jobs as there are jobs in agriculture itself. If this is still the case it suggests that between 4% and 4.5% of the labour force in 1995 were in jobs which would be affected by the expansion or contraction of the agricultural industry. What happens in times of agricultural contraction is that the demand for fertilisers decreases, for example, so that fertiliser firms question the need for sales representatives and make some of them redundant while, if less milk is produced, fewer workers are required to process it into butter and cheese and milk processing factories are closed down.

Table 8.1   Whole-time employment related to UK agriculture in 1981.

Employment sector	Agriculture-related jobs (whole-time equivalents)	% of UK whole-time equivalents
Agriculture and horticulture	524 187	2.44
Input industries	298 527	0.93
Public sector	40 000	0.19
Output industries	244 037	1.14
Total	1 006 751	4.69
Total in all employment	21 449 641	100.00

Source: Andrew Errington (1986) 'Employment', in G.M. Craig, J.L. Jollans and A. Korbey (eds) *The Case for Agriculture: An Independent Assessment*, p. 98, CAS Report 10, Reading.

The balance of payments

The UK often has a balance of payments deficit (see Table 8.2), so it is argued that producing food at home would help to decrease imports and so reduce the size of the deficit.

Table 8.2 The UK balance of payments.

Year	Current balance (£million)*
1982	4 649
1983	3 529
1984	1 482
1985	2 238
1986	−871
1987	−4 813
1988	−16 475
1989	−22 398
1990	−19 293
1991	−8 533
1992	−9 468
1993	−11 042
1994	−2 080
1995	−6 670

Source: Office for National Statistics (1996) *Monthly Digest of Statistics*, p. 100, May.
* A negative figure indicates a balance of payments deficit.

While this argument is not incorrect, it may be simplistic. There are other ways, apart from import saving, of solving a balance of payments problem. One is to devalue the currency, and so make imports more expensive, so that fewer are bought, and another is to increase taxes or unemployment, so that consumers feel poorer and so buy fewer imported goods. A study carried out in the 1970s suggested that import saving was the best of the three methods, but it did not conclude that agriculture was necessarily the best industry to choose[2]. Presumably the best industry to choose would be the one which produced the greatest import saving for the smallest investment, and there is no certainty that this would necessarily be agriculture. Indeed, because the proportion of agricultural products in the total import bill has decreased over the last quarter century, while the proportion of manufactured goods has increased (see Table 8.3), it may be that replacing imports of manu-factured goods would have a bigger impact than replacing agricultural imports. At the same time, it must be remembered that the trends illustrated in Table 8.3 have occurred in the presence of farm income

Table 8.3   The composition of UK imports.

	Food and live animals (% of total)	Manufactured goods (% of total)
1970	20.6	50.6
1990	8.3	77.8
1994	8.3	81.4

Source: Calculated from CSO (various editions) *Annual Abstract of Statistics*, HMSO, London.

support by protectionism: were this to be withdrawn, the proportion of agricultural imports would presumably increase again.

The other question that has to be answered in the context of import substitution is 'which imports should be substituted?'. At present, in a protectionist CAP, the only products in which the UK is not close to self-sufficiency are oilseeds, sugar, fruit and vegetables, bacon and ham, and some dairy products (see Table 7.5). With free trade, cereals and meat would presumably be imported to a greater degree, as they were before the UK became a member of the EU, and especially before World War II.

The arguments *against* greater self-sufficiency

Comparative advantage

If you can't remember what this means, go back to the appendix in Chapter 7. The theory of comparative advantage suggests that countries should export the products in which they have a comparative advantage, and import the products in which they do not have a comparative advantage. The problem is to decide which is which. One method is to examine production costs: the country with the lower costs has the comparative advantage. The problem with this is that farmers receiving higher prices will be prepared to spend more on inputs, so we do not really know what UK or EU production costs would be if UK or EU farmers were receiving the same price as farmers in the USA or New Zealand. It seems likely that the USA might have a comparative advantage in cereal production, and New Zealand in sheepmeat and dairy products, but it is difficult to prove conclusively. On the other hand, we do know that within the EU, where prices are roughly similar, the UK is likely to have a comparative advantage in grass-based products over countries like France and Italy.

But the problem is not confined to agriculture. Traditionally, the UK had a comparative advantage in manufactured products and services, but whether this remains the case for the UK as a member of the EU is more difficult to judge. Thus, the answer to the comparative advantage question depends upon the way the question is specified. Are we concerned with the UK or with the EU, since agricultural support questions are decided within the context of the CAP, and are we talking about trade within the EU or with the rest of the world?

Retaliation

This, in contrast, appears to be a much simpler issue. If we do not buy goods from producers in other countries, we decrease their ability to buy goods from us. But again there are complications. What is to stop New Zealanders from selling sheepmeat to the EU and using the foreign exchange so earned to buy Japanese cars? Nothing at all. This has led some people to argue that the possibility of retaliation should be discounted, but others argue that, in this example, extra sales by Japanese car workers increase their ability to buy Scotch whisky or French wine. Thus, the benefits of free trade, its proponents argue, eventually rub off on everybody and this is the principle underlying the General Agreement on Tariffs and Trade (GATT).

GATT

The simplest argument against greater self-sufficiency is that the UK and the EU have obligations under GATT agreements to allow a certain level of imports (see Chapter 7).

⇨ **Questions**_____

Now review your understanding of the trade arguments for farm income support by attempting the following:

1 If the UK imports more tropical fruits such as mangoes, guavas and passion fruit, what effect will this have on its self-sufficiency in indigenous products?
2 Is there any economic reason why a country should attempt to be 100% self-sufficient in a commodity?
3 What is meant by the term 'Balance of Payments'?
4 Can you now see why it might be interesting to define self-sufficiency at world price levels, or by specifying the use of indigenous inputs only?

8.3   The income arguments

The arguments about farm incomes can be divided into two parts:

- Do farmers really have low incomes?
- If they do, should anything be done about it?

Question 1: do farmers really have low incomes?

What do we mean by low incomes? Lower than they used to be? Lower than those of comparable businesses (and, if so, what are comparable businesses?)? Or low in relation to the amount of capital invested? And what, precisely, is farm income? Some of these questions are easier to answer than others.

What is farm income?

At first sight this is one of the easy questions. It's presumably the same as the farmer's profit, which is the difference between what is bought and what is sold. Yes, but then this figure has to be adjusted for depreciation of buildings and machinery, interest payments, rent and labour costs, before you get the figure which the farmer and his or her spouse receive as income. This is known in the official MAFF figures as *Farming Income* (see section 6.8). But then, of course, the value of money changes from year to year, so to get an estimate of the real income of farmers we need to adjust for inflation, which is what has been done in Table 8.4.

Table 8.4   Index of farming income in the UK at constant prices (average of 1940–69 = 100).

Year	Index	Year	Index
1975	95	1985	37
1976	113	1986	46
1977	98	1987	50
1978	89	1988	34
1979	77	1989	45
1980	56	1990	41
1981	66	1991	36
1982	77	1992	49
1983	57	1993	77
1984	86	1994	81 (forecast)

Source: J. Nix (1995) *Farm Management Pocketbook*, 26th edn., p. 204, Wye College Press, Ashford.

Table 8.4 suggests that farm income is decreasing, with lots of ups and downs, but of course we should remember that these figures are for an agricultural industry which is already being supported. These figures do not tell us what would happen in the absence of support, and we should remember that we are trying to decide the question of whether or not that support should be given at all. Therefore, presumably, we should deduct expenditure on intervention, headage payments, capital grants, Hill Livestock Compensatory Allowances and all the other forms of government expenditure, which in the financial year ending in March 1996 was forecast to be £3033 million. Since Farming Income for 1995 was estimated at a little over £4000 million, it appears that without government money the agricultural industry would have made only about £1000 million in that year. Of course, we should remember that, in the absence of support, prices would be different and the number of farmers would probably be different, so we can only look upon the figures produced by this calculation as a very rough estimate of what would happen.

Are farm incomes lower than they used to be?

Table 8.4 suggests that they are lower than they were in the 1950s and 1960s. Longer term comparisons suggest that farm income levels for most of the years in the 1980s may have been lower than they were in the 1920s and 1930s when there was little government support[3].

Are they lower than they are in comparable businesses, or in relation to the amount of capital invested?

It all depends on what a comparable business to farming is like. Do we mean other small businesses, such as corner shops, or other rural businesses, such as forestry, and are these comparisons meaningful? It is easier to say something about agricultural incomes in relation to the amount of capital invested, because we know that it is not very high, especially in relation to the amount of capital invested in land.

So if we look at farm incomes, strictly defined, it might be argued that they are not very high. But many farmers have incomes from non-farming activities, and many owner-occupying farmers have been made wealthier by the increase in the value of their land. Can we really argue that a farmer who has assets of £250 000 needs income support? Yet that would be the capital value of only a small farm. Would it really produce a big enough income to support a family?

If farm incomes are low, why do farmers stay in agriculture, when they might earn more by putting their money in the bank and living off the interest? Part of the reason might lie in the observation that many farmers appear to find agriculture a pleasant way of life, being their own boss in a beautiful landscape (the technical term which is used for this is *psychic* income). On the other hand, you can't eat the scenery and suicide rates among farmers are said to be higher than in other occupations.

Question 2: if farmers do have low incomes through no fault of their own, should anything be done about it?

There are other groups in society who also have low incomes through no fault of their own, and receive grants or income support from the state. They are students, old-age pensioners, the mentally and physically disabled and the unemployed.

However, there are other groups in society who also have low incomes, or who work in traditional and declining industries, who receive little or no state assistance. Examples are small shopkeepers, coalminers and steelworkers.

Put like this, it becomes clear that when we ask if farm incomes should be supported just because they are low we are examining a question of value judgement, about which economics has nothing useful to say. Society normally answers questions about value judgements through the political mechanism and, clearly, the function of political pressure groups such as the National Farmers' Union, the Country Landowners' Association, the farm workers' union (part of the TGWU) and their equivalents in the EU and in other countries is to convince governments that there are good reasons for supporting farm incomes. In order to do so, they will use some of the other arguments we have been discussing in this chapter.

⇨ **Questions**_____

Now see if you can answer the following on the income arguments:

5 What are the main reasons for the fluctuations in farm incomes observable in Table 8.4?
6 Why might you expect that in the long term, in the absence of farm income support, farm incomes would fall?

8.4   Farm income support and the rural economy and society

This part of the discussion is related to the question of the employment effects touched on under the trade arguments, but it goes further because it implies that if agriculture has problems, then so will the rural economy, and therefore so will rural society. There are several assumptions built into this last sentence which we should take care to make explicit:

- Will problems for agriculture mean problems for the rural economy?
- Will problems for the rural economy produce problems for rural society?
- If rural society has problems, does it matter?

Let us take first the question of whether problems for agriculture affect the rural economy. The answer appears to be obvious. Agriculture is the main rural industry, and so problems for agriculture must mean problems for the rural economy because agriculture dominates rural employment. But is this still true? There are now fewer than 100 000 full time hired agricultural workers, and less than 500 000 people with full-time farm jobs in the UK. Agriculture only accounts for 2.2% of the workforce in employment, but we would expect it to account for a much larger proportion of the *rural* labour force. Nevertheless, Hill and Ray estimated that it only accounted for about 14% of the rural labour force in the 1970s and, by 1989, the figure was down to 3% in accessible rural areas (those within commuting range of towns) and 6.2% in remote rural areas[4]. Recreation, tourism and light industry are becoming increasingly important in those areas which we can still define as rural. It should be noted, however, that this conclusion applies to the UK; it may not apply to the same degree, if at all, to other parts of the EU.

Even if agriculture dominated the rural economy, it would not necessarily follow that problems for the rural economy would produce problems for rural society, for, with the advent of the motor car and the possibility of commuting, people who live in the countryside no longer necessarily work there. There can be few areas of lowland Britain that are outside commuting range of a major location of urban jobs, so it is quite possible to have a run-down rural economy in many areas without producing high levels of unemployment simply because rural people commute to work. This is perhaps the main reason why the decrease in the rural population of the UK that took place until the early 1960s was halted by the 1971 census and quite clearly in reverse from the 1981 census onwards. In France, in contrast, the rural population has declined more or less steadily since the 1860s[5]. The effect of this population renaissance on rural society is controversial. Some argue that commuters do not make the same contribution to village society as those who work in the countryside; others contend that they take over village institutions and organisations from those who traditionally ran them. It is questionable whether both views are tenable simultaneously.

And finally, there is the issue of whether or not it *matters* that rural society might have problems. This is largely a matter of value judgement, and therefore a political question, again one on which economists have no useful comments to make. As members of society they may regret that any part of that society should suffer difficulties, but the same would presumably apply to inner-city problems.

⇨ **Questions**_____

Now consider the following:

7 If we are concerned with the rural economy and society, presumably we have to define a rural area. How would you define one?
8 Why might commuting be more common in rural England than in highland Scotland or south western France?

8.5 The environmental arguments

There are those who argue that Britain is still beautiful, so there is no environmental problem associated with agriculture, but, given the data in Table 8.5, this is not an easy view to sustain.

Table 8.5 Habitat loss in Great Britain, early 1950s to early 1980s.

Habitat	Percentage loss
Lowland herb-rich grassland	95
Chalk and limestone grassland	80
Lowland health	60
Limestone pavement	45
Ancient woodland	50
Lowland fens and marshes	50
Lowland raised bogs	60
Upland grassland, heaths and mires	33

Source: P. Lowe *et al.* (1986) *Countryside Conflicts*, p. 55, Gower, Aldershot.

Clearly, not all of these changes are the result of agricultural activities, but many of them are. Indeed, these are only the changes which have affected wildlife conservation. There have also been changes which have affected the appearance of the landscape, such as the demolition of old buildings and their replacement by new ones, which many people regard as a retrograde step. In fact, what is good for wildlife is not necessarily good for landscape. Fields which are not grazed will probably revert to scrub, which may be quite good for some forms of wildlife but not to the taste of a population which appears to appreciate a farmed landscape.

The pro-income support argument is that farmers will not be interested in conservation if they are not making money, but will happily plant trees and maintain buildings if they are feeling rich. The usual argument against this is that a few young trees are no replacement for the mature trees and mature habitats which are destroyed when farmers, stimulated by high prices, reclaim land for agricultural purposes in order to increase output. Similarly, it is argued that the effect of high farm incomes is to stimulate the removal of hedges and investment in new farm buildings which disfigure the landscape rather than the maintenance of hedges and the repair of old farm buildings which are an integral part of it.

8.6   Summary

Could you conclude

- from the trade arguments,
- and the income arguments,
- and the arguments about the rural economy and society,
- and the conservation arguments,

that we should or should not support farm incomes?

The most obvious conclusion is that the arguments do not allow a clear-cut decision on one side or the other. In which case it is interesting to find that all industrialised countries, with the present exception of New Zealand, *do* in fact have some form of support[6].

Notes

1. Richard Body (1984) *Farming in the Clouds*, pp. 77–78, Temple Smith, Hounslow.
2. B.Hill and K.Ingersent (1984) *An Economic Analysis of Agriculture*, 2nd edn., p. 184, Heinemann, London.
3. There are estimates of farming income back to 1938 in H.F. Marks and D.K. Britton (1984) *A Hundred Years of British Food and Farming*, p. 150, Taylor and Francis. The book by E.M. Ojala (1952) *Agriculture and Economic Progress*, p. 215, Oxford University Press, Oxford gives agricultural income estimates from 1867 to 1939 which are based on a similar but not exactly comparable method of calculation. These suggest that farming

incomes in the 1980s were generally below the levels of the inter-war years, and that the recovery of farm incomes in the 1990s brought them back to inter-war levels in real terms.

4. B.Hill and D.Ray (1987) *Economics for Agriculture*, p. 416, Macmillan, London; Secretary of State for the Environment and Minister of Agriculture Fisheries and Food (1995) *Rural England: A Nation Commited to a Living Countryside*, White Paper (Cm 3016) p. 30.

5. A.Moulin (1991) *Peasantry and Society in France Since 1789*, Cambridge University Press, Cambridge; and for evidence on decreases in agricultural employment in other EU countries see Eurostat (1994) *Agriculture: Statistical Yearbook, 1994*, p. 24, Brussels.

6. You can find a totally different way of arguing these issues, from a political and spiritual viewpoint, in M.Allaby and P.Bunyard (1980) *The Politics of Self-Sufficiency*, Oxford University Press, Oxford.

Chapter 9
The Historical Background to Agricultural Policy

9.1 Introduction

The previous chapter demonstrated that there is no indisputable economic case for supporting farm incomes. Yet most industrialised countries do support them. This section of the course is therefore concerned with the historical and political reasons why.

We shall also try to account for the significant differences in the structure of agriculture between the UK and other member states of the EU that we have previously observed.

9.2 Before World War II

Over much of western Europe in the medieval period, farms were small and their land was scattered in small parcels around common fields. By the middle of the nineteenth century, farms in Britain had been enclosed so that they were no longer worked in common and had grown in size, as landlords had arranged their estates into relatively large farms leased to capitalist tenants. The population was expanding rapidly but industrialisation had created a demand for labour, so many of the extra hands and mouths had moved out of the countryside to the towns and cities. Those who were left worked in an increasingly commercialised agricultural industry producing food for the urban market.

In the UK, farm incomes were supported to some extent by the Corn Laws, which imposed import duties on wheat and other cereals except when prices were very high, until their repeal in 1846. Thereafter, there was no support until 1939, with the exception of the World War I years and some subsidies in the 1930s. Opponents of Corn Law repeal had predicted that it would result in the immediate and rapid demise of British agriculture. But the Crimean war interrupted trade from Russia, then a major exporter, in the 1850s, and the Civil War prevented the rapid expansion of the railways in the United States in the 1860s. Thus, in the event, a series of unforeseen developments kept large volumes of imports out of the country for 30 years, and the period between 1850 and 1875 was later described as a 'Golden Age' of British agriculture. It was followed by what cereal producers felt was a great depression. The westward progress of the railways into the Great Plains of the USA, coupled with the introduction of steam engines into ocean-going ships, reduced the cost of transporting grain from the interior to the eastern seaboard, and from there to Europe. Table 9.1 shows the effect that this had on prices.

Table 9.1   Freight charges and wheat prices (pence per quarter).

	1870–74	1880–84	1895–9
Freight charges			
Chicago–New York, by rail	113	63	47
New York–Liverpool, by steamship	66	35	23
USA wheat price at Liverpool	625	531	356

Source: M. Tracy (1982) *Agriculture in Western Europe: Challenge and Response, 1880–1980*, 2nd edn., p. 20.

In the UK, the effects were initially felt by cereal producers, but shipboard refrigeration became available from the late 1870s and imports of meat and dairy products began to increase, not only from the USA but also from Australia and New Zealand. Between the 1870s and 1890s, wheat prices halved, the prices of barley and oats fell by about a third, and cheese, butter, cattle and sheep prices fell by about a quarter. By the beginning of the twentieth century, the UK was importing three-quarters of its wheat and cheese, about half of its beef, lamb and barley, and most of its butter. In the 1930s, domestic self-sufficiency stood at about 33% for all foods and 50% for indigenous foods (all these figures are taken from D.Grigg (1989) *English Agriculture: An Historical Perspective*, pp.9, 22, Blackwell, Oxford). By the beginning of World War II, the UK was the biggest importer of agricultural products in the world.

Continental Europe, in contrast, maintained a much higher level of self-sufficiency. The process of transforming the common European medieval pattern had produced variable results by the end of the eighteenth century, with technically advanced and intensive farms in the Low Countries contrasting with large landlord-run grain-producing estates in Prussia and small peasant holdings in France, Italy and southern Germany. The upheaval resulting from the French Revolution after 1789 produced the abolition of feudal privileges in France, as the peasants became owners of their small, scattered holdings, and the laws embodied in the Code Napoleon maintained the practice of division of land among all heirs. Thus, if the population increased, what had begun as small farms became smaller still. More or less the same thing happened in many of the areas taken over during the expansion of Napoleonic France. Industrialisation also took place later in these areas so that over much of western Europe, the countryside in the mid-nineteenth century was populated by a landowning peasantry with little incentive to leave their farms and no great industrial demand for their labour.

Following the abolition of import duties by the UK, many continental European countries also abolished or at least reduced import control: Germany in 1853, France in 1860, and Belgium in 1862, for example. Then, as exports from the new world increased and prices fell, one country after another reintroduced protection: Germany, in a series of measures between 1879 and 1902; Italy in 1878, with increased rates of duty in 1887; France, in 1881 and with higher tariff rates in 1885, 1887 and 1892. The main exceptions to the trend were Denmark, which responded to cheaper grain by transforming its agriculture from an emphasis on cereals to dairy and pig production, and the Netherlands, which concentrated on dairy products, fruit and vegetables. The pattern was maintained after World War I, with an even greater emphasis on national self-sufficiency in the 1930s in the fascist economies of Germany and Italy. It was also the depression years of the 1930s which provoked the first examples of farm support in the exporting countries, with the establishment of the Agricultural Adjustment Act of 1933 in the USA, the Wheat Board in Canada in 1933 and subsidised exports from Australia in 1938 and New Zealand in 1936.

These contrasts between Britain and continental Europe are summarised in Table 9.2, which takes the story up to 1939. The important point to notice is that the results were different: by 1939, the UK was the major world importer and the continental European countries were either aiming at self-sufficiency or had become exporters themselves.

⇨ **Questions**_____

Now test your understanding of the pre-1939 period by answering the following:

1 Why did the price of American grain fall after the 1870s?
2 What is the main structural difference between UK and continental European agriculture?
3 Canada, Australia, and New Zealand were British colonies in the late nineteenth century, the USA had been a colony, and Argentina had

many people of British origin associated with its agricultural industry. What difference do you think this made to attitudes to imports of food *in Britain*?

Table 9.2   Contrasts in agricultural policy 1846–1939.

The UK	Continental Europe
Big farms, because of the way in which medieval peasants had been transformed into capitalist farmers renting land from landlords and employing labour	Small farms, because of the way in which medieval peasants had taken over their farms, together with inheritance laws which split these farms between all their children
plus	plus
early industrialisation, attracting workers out of agriculture,	late industrialisation
led to	led to
a small agricultural labour force by the mid-nineteenth century,	a large agricultural labour force
and, coincidentally,	and, coincidentally,
British colonies were major temperate product producers	colonies of most continental countries produced mostly tropical products
so that	so that
from the mid-nineteenth century Britain adopted free trade, which meant that from the 1870s she became a food importer on a gradually increasing scale.	when cheap grain became available from the new world, most continental countries remained self-sufficient in food, and Holland and Denmark became exporters.

9.3   After World War II

The UK's reliance on imports and the free market, which was beginning to be challenged even before the outbreak of World War II, came to a rapid halt in 1939. Now that cargoes of Australian wheat, New Zealand lamb and Argentinian beef could be sent to the bottom by U-boats and surface raiders, and shipping space was needed to bring tanks and guns and soldiers from America, the emphasis in Britain was on feeding the country from its own resources – 'digging for victory' in the phrase of the time. The government provided guaranteed prices to persuade farmers to make the necessary investment in buildings, machinery and land reclamation and, partly as a result of this and partly by strict rationing, especially of meat, butter, cheese, eggs and sugar, food supplies were maintained.

At the end of the war, therefore, the new Labour government had to decide whether or not it would return to a policy of free trade, or whether it would maintain the wartime policy of using the price mechanism to promote home production. There were conflicting considerations to take into account:

- The war had demonstrated that UK food consumers had been heavily dependent on imported food, and that the trade routes along which it came were prone to disruption by enemy action.
- There was perceived to be a rural poverty problem, and farmers were thought to be deserving of support for their hard work during the war.
- On the other hand, it was predicted that the immediate post-war world food shortage would rapidly be replaced by increased supplies (since wartime production capacity was still available and ships were no longer being sunk) and lower prices, as had happened at the end of World War I.

Thus, the government took the view that it would like to be in a position to take advantage of lower world prices if and when they came, while at the same time maintaining an agriculture with the capacity to expand rapidly should the need to do so arise. The much-quoted preamble to the 1947 Agriculture Act stated that its objective was to promote and maintain

'... a stable and efficient agricultural industry capable of providing such part of the nation's food and other agricultural produce as in the national interest it is desirable to produce in the United Kingdom, and of producing it at minimum prices consistent with proper remuneration and living conditions for farmers and workers in agriculture and an adequate return on the capital invested in the industry.'

The main policy mechanism designed to achieve this objective, to take advantage of any available cheap food while simultaneously maintaining a domestic industry with a capacity for rapid expansion, was the *deficiency payments system*:

- Consumers bought food, whether imported or home produced, at world prices.
- A guaranteed price was set by annual negotiation between the government and the farmers' unions.
- Market prices were assessed each week, and farmers received a deficiency payment, a cheque (financed by the taxpayer) for the

difference between the average market price in the week in which their products were sold and the guaranteed price.

In continental Europe at the end of World War II, things were different. Most countries:

- had few existing trade links with major exporters;
- still had a high proportion of their labour forces on the land;
- recently had their economies devastated by the war, and so were producing very little which they could export to pay for imported food.

The choice was simple: they re-erected their protectionist barriers and tried to be as self-sufficient as possible. Clearly, there were differences in detail from one country to another, but this was the basic pattern. It meant that when the new European Community was established at the beginning of 1958, and it was decided that an agricultural policy common to all member states was necessary, the task was relatively easy because there was some basic similarity between their existing policies. All they had to do was to place the protectionist barrier at the Community frontier rather than at national frontiers, and they had a Common Agricultural Policy (CAP). The contrasts with the UK are summarised in Table 9.3.

Table 9.3   A brutally simple view of contrasts in agricultural policy, 1945–73.

The UK	Continental Europe
The UK wanted to continue to take advantage of any cheap food which was available on the world market, while supporting domestic agriculture so as to maintain some supply security	At the end of World War II, continental countries, not having any money to import food with anyway, continued with their protectionist policies and high levels of domestic self supply
which led to	which led to
the deficiency payments system, which lasted until the UK joined the European Community, when the UK also joined	the decision, when they formed the EC, to combine their already-similar protectionist farm policies and call the result
the CAP	the CAP

Other developed countries have also supported their agricultural industries in the post-war period. The USA has had a complex of support mechanisms, including measures for supporting domestic prices, acreage reduction programmes for supply reduction and conservation purposes, consumer support in the form of Food Stamps, and export subsidies. In recent years, as the number of farmers has fallen and environmental concerns have increased, successive administrations have attempted to reduce the level of support to farmers while attempting to raise market prices by increasing world price levels. This explains the US position in the GATT negotiations (discussed in the previous chapter), and the degree of its co-operation in those negotiations with other traditional exporting countries such as Canada, Australia and New Zealand. These last three countries have had farm income support programmes, although they have perhaps been more concerned with the stabilisation of export earnings for their farmers. Since 1985, New Zealand has significantly reduced the level of farm income support, as its 1992 PSE figure of 3% suggests (see Table 7.3).

In contrast to these traditional food exporters, Japan, as a densely populated industrialised country, appears to be a natural food importer. However, the supply security argument has proved very powerful in Japanese agricultural politics, to the point where the price of rice, which is not only a staple food but has a quasi-religious significance, is heavily supported. At one point the price to producers was eight times the world price level. On the other hand, imports of other agricultural products have significantly increased.

To summarise, you can see that most of the industrialised countries of the world have supported their agricultural industries to a greater or lesser extent over most of the twentieth century (see Table 9.4). This in itself helps to account for the continuation of support, since there is often a kind of political inertia which helps to prevent rapid and radical change. There are also other, political, reasons, which we shall examine below.

Table 9.4 OECD estimates of transfers of money associated with agricultural policies.

	Total transfers 1992 (billion ECU)	Total transfers in 1992 per full-time farmer equivalent (ECU)
Australia	1.2	3 200
Canada	3.3	15 800
EU–12	120.5	13 700
Japan	57.2	18 500
New Zealand	0.0	300
USA	70.4	27 900

Source: OECD (1993) *Agricultural Policies, Markets and Trade*, pp. 160, 162, OECD, Paris.

⇨ **Questions**_____

Now see if you can answer the following about post-World War II agricultural policies:

4 What do you think were the advantages and disadvantages of the UK's deficiency payments system?
5 Was post-war growth in farm output the result of agricultural policy or technical change? (Think about the arguments on either side and don't worry if you can't come to a definite conclusion.)
6 Why did the Irish Republic and Denmark find it advantageous to join the EC at the same time as the UK?

9.4 The politics of agricultural policy

We have seen that we cannot explain the decision to support farm incomes on purely economic grounds, and that tradition has something to do with it. But there is probably more to it than that, because tradition by itself would be unlikely to persuade politicians to spend the funds that they do spend on farm support. We are therefore left with the conclusion that farm lobbyists must be doing what they are paid to do. The difficult thing to explain is why an industry which accounts for such a small proportion of the output of industrialised countries, and employs such a small proportion of the labour force, should be so effective politically.

The way in which decisions about agricultural policy are made in the EU is discussed in the next chapter. You will see then that one of the main conclusions which emerges from that discussion is that *pressure groups* are influential in the whole process, both at a national level and in the EU as a whole.

But this only tells us that pressure groups are influential. It does not explain why the pressure groups which argue for increased farm support should be more influential than those which argue for less. However, there are several reasons for suggesting that farm pressure groups (the whole lot together are sometimes called the *farming lobby*) will be more successful in arguing their case than groups representing food consumers (and so wanting lower prices) or environmentalists (who want fewer incentives for farmers to increase output). In the UK these reasons are:

● Producer groups often have more money than consumer and environmental groups because they find it easier to persuade potential members of the advantages of membership. Put crudely, a

10% change in farm prices will produce a 10% change in farm incomes, but if consumers are only spending less than 20% of their incomes on food, it will affect their expenditure by less than 2%. So farmers and those who sell inputs to them will have good reasons for joining the pressure groups representing their industry. Consumer and environment groups will have to spend part of their income and energy on retaining existing members and recruiting new ones. The result is that the number of people employed by the National Farmers' Union (NFU) and other producer organisations to put the case for farmers far exceeds the number of people employed by their opponents to put the opposite case. And the more people you employ, the more politicians, civil servants and journalists there are to whom you can talk.

- The agricultural ancillary industries – those producing fertilisers, pesticides, farm machinery and a whole host of other products down to baler twine and barbed wire – know that high farm prices produce higher sales than low farm prices, so groups like the Fertilizer Manufacturers Federation will tend to support the arguments of the farmers' own organisations.

- In the UK, the NFU and the ancillary industries have developed a close relationship with the Ministry of Agriculture, Fisheries and Food (MAFF), at both the politician level – the Minister of Agriculture is said to see the President of the NFU at least once a week – and the official level, so that NFU officials have many opportunities to speak to their opposite numbers in MAFF. In part, the reason for this goes back to the days of deficiency payments, when the NFU had a statutory right to be consulted over agricultural policy. Although this is no longer the case, the feeling still remains in MAFF that its role is to look after the interests of the agricultural industry. The consumer and environmental groups have not been successful in developing a similar relationship with any corresponding government department, partly because they lack the resources to do so and partly because they have simply entered the game more recently.

Thus, there are good reasons why farmers will be successful in getting their views over to politicians, civil servants and journalists in the UK. Obviously, they do not have total control over the views of the decision-makers, or of the people who elect them, and in recent years it has been possible to discern a change in public attitudes to agriculture. At the end of World War II, farmers were popularly perceived to have done a good job in feeding the country and were seen as the people who looked after the countryside. From the 1970s onwards, in the face of increasing surpluses of farm produce and mounting evidence of the damage to wildlife and the landscape caused by modern farming methods (see Chapter 8), popular attitudes towards agriculture began to change. Nevertheless, it is possible to argue that these changes in attitude would have had a much greater effect on farm policy than they did, had the farming lobby not been so effective in putting the case for agriculture.

There are similar groups to the NFU in other member states of the EU. In Germany its equivalent is the Deutscher Bauernverband (DBV), which has links with a political party, the Christian Democrats, and its associated Christian Social Union (CSU). One of the more influential agriculture ministers in recent years, Ignaz Kiechle, as well as being a Bavarian farmer, was also one of the leaders of the CSU. Before him, when a coalition of the Socialist party and the Free Democrats was in power, another Bavarian farmer, Joseph Ertl, was one of the leaders of the Free Democrats and the Minister of Agriculture. Thus, whatever lobbying power the DBV has had has been strengthened by the political importance of the minister in the ruling coalition, which helps to explain why such an industrialised country has so firmly resisted any move to cut farm support levels within the EU.

In France, the nearest equivalent to the NFU is the Federation Nationale des Syndicats des Exploitants Agricoles (FNSEA), but there are other organisations too. Some younger, reformist farmers belong to the Centre National des Jeunes Agriculteurs (CNJA) and, in 1987, several peasant farmers' organisations joined together to form the Confederation Paysanne. The differences in emphasis between these various groups reflect the differences in structure and objectives between various sectors of French farming, with big commercial farms in the Paris Basin interested in increasing production and developing export markets, contrasting with the peasant holdings of the west and south-west mostly concerned with maintaining the rural way of life. It is perhaps this lack of a single political focus which helps to explain why disagreements over policy issues can sometimes result in violent demonstrations in the streets of France.

The function of farmer organisations is to persuade the governments in individual member states to espouse the policies which they believe will benefit their members. Thus, the NFU in the UK will be more concerned with the welfare of bigger farmers at the expense perhaps of smaller farmers in the UK. The equivalent of the NFU in Greece, Spain, Portugal and Italy will emphasise the needs of small producers in undeveloped areas. Dutch, Danish and Irish farm organisations will be concerned with exports to other member states, and so on. But in addition to attempting to influence their own national governments, they will also attempt to influence the European Commission. The Commission prefers to be lobbied by EU-wide organisations, rather than national bodies, so all the farmers' organisations join together to form the Comite des Organisations Professionelles Agricole (COPA). There are also equivalent organisations for other interests: COCEGA represents agricultural co-operatives, BEUC (the Bureau Europeene des Unions des Consommateurs) represents consumers, and the EEB (the European Environmental Bureau) puts the case for the various EU environmental organisations. These are the bigger bodies, but many trade associations have representatives in Brussels to put their case to the Commission – the European Association of Manufacturers of Fruits and Vegetables preserved in Vinegar, Pickling Brine and Oil is just one precisely defined organisation as an example.

We shall return to the question of decision making about agricultural policy in the next chapter. The purpose of this chapter was to explain why industrialised countries support farm incomes even though there is no clear or generally accepted economic case for doing so. When arguments are complex, tradition, coupled with effective farm lobbies, can exert a powerful influence and provide as good an explanation of the decisions that have been taken as any analysis of economic forces.

⇨ **Questions**_____

Now use these questions to review your understanding of the discussion on the politics of agriculture:

7 Can you identify the UK pressure groups which look after the interests of landowners and farm workers in the same way that the NFU looks after the interests of farmers?

8 List three reasons for the effectiveness of the farm lobby in the UK.

9 Can you suggest why the UK farm lobby might not be so influential as it once was?

10 Would you expect COPA to be as effective in the EU as national farm organisations might be in individual member states?

9.5 Summary

- Before World War II, Britain had good reasons for not supporting farm incomes, while the continental European countries had equally powerful reasons for being protectionist.
- After World War II, the UK developed a system of agricultural support which combined access to the world market with maintenance of farm incomes.

- The CAP began as a logical development of the existing protectionist agricultural policies of the original member states of the EU.
- The political power of the agricultural industry helps to explain why industrialised countries, with only a small proportion of their populations in farming, nevertheless spend large amounts of money on farm income support.

Chapter 10
The CAP and How It Works

10.1 Introduction

In detail, the Common Agricultural Policy of the European Union (to use its full title) is now so complex that there is probably no one person in the whole world who understands completely how every bit of it works; but in essence, its basic outline is so simple and logical that you can understand it quite quickly. That's as far as most people need to go, and it's what this chapter is about. It is divided into several parts:

- The need for a *common* agricultural policy
- The origins of the CAP
- CAP reform
- An outline of the policy mechanisms
- An outline of decision-making mechanisms in the EU
- The future of the CAP

10.2 The need for a *common* agricultural policy

We have already seen, in earlier chapters, that agricultural industries in developed countries face fluctuating and declining prices, and that in many cases the governments of such countries decide to introduce some form of agricultural policy to combat what they perceive to be the resulting problems. By the middle of the 1950s, the governments of the original six members of the EU (see Table 10.1) had adopted agricultural policies which were based on a system of import controls and price support. These kept prices higher than they would have been in an unsupported market, and so maintained agricultural incomes. There were general similarities between these policies but numerous differences in detail.

When the six countries negotiated among themselves to set up a common market in which barriers to trade would, as far as possible, be eliminated, it was agreed that the exclusion of agriculture from this general common market would be impossible. Without free trade in farm products, national food price levels might differ, and those countries with lower food prices could offer lower wages and so have lower industrial costs. This would undermine the effects of the free-trade policies which would apply to other products. Consequently an agricultural policy which applied to all member states was required – a *common* agricultural policy.

Table 10.1   The member states of the European Union.

Country	Date of accession
Belgium, France, Germany, Italy, Luxembourg, the Netherlands	1958
Denmark, Ireland, the UK	1973
Greece	1981
Portugal, Spain	1986
Austria, Finland, Sweden	1995

10.3   The origins and development of the CAP

Articles 38–47 of the Treaty of Rome (the treaty, signed in 1957, which established the European Economic Community, as it then was) apply directly to agriculture. The objectives of the CAP were stated in article 39.1:

(1) To increase agricultural productivity by promoting technical progress and by ensuring the rational development of agricultural production and the optimum utilisation of the factors of production, in particular labour;
(2) thus, to ensure a fair standard of living for the agricultural community, in particular by increasing the individual earnings of persons engaged in agriculture;
(3) to stabilise markets;
(4) to ensure the availability of supplies;
(5) to ensure that supplies reach consumers at reasonable prices.

Subsequent articles lay down the guidelines for the policy instruments by which these objectives are to be achieved and the arrangements for establishing the detailed provisions of the CAP. These emerged in the early 1960s, although the three principles which are often said to be fundamental to the CAP system – the single market, joint financing and Community preference – do not seem to have appeared in any formal document. They emerged early in the working life of the CAP, which began when the first products were subjected to a common set of market rules in 1962. Common prices were first applied in 1968.

Under the terms of this common policy, the price received by producers was determined by market forces, but these market forces were controlled so that prices could only fluctuate between upper and lower limits. If they rose above the upper limit it became worthwhile for traders to import supplies from outside the EU; if they fell below the lower limit it became worthwhile for traders to sell into *intervention stores* or, with the aid of *export refunds*, on to the world market. These arrangements were applied, with few basic changes, until the middle of the 1980s. At the same time technical changes in agriculture (new varieties, more fertilisers and pesticides – see Chapter 5 on the supply of agricultural products) meant that yields rose by about 2% per year for the EU as a whole, while demand was generally fairly static. The result of this is shown in Fig.10.1, in which the effect of the yield increase is to shift the supply curve from S_1 to S_2.

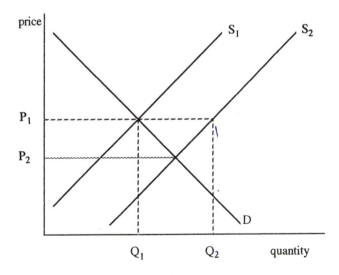

Fig. 10.1   The effect of the CAP on prices.

Fig. 10.1 demonstrates that, although in an unsupported market prices would have fallen from P_1 to P_2, the effect of the CAP was to maintain prices at P_1 with the excess quantity supplied $(Q_2 - Q_1)$ going into intervention stores or on to the world market with the aid of export refunds. By the middle of the 1980s, it was clear that the CAP was producing several problems:

(1) The EU was producing surpluses of farm products which were expensive to store and expensive to sell on the world market. Expenditure on agriculture was accounting for about 60% of the EU budget and had doubled compared with the late 1960s, even after allowing for inflation and the entry of additional member states.

(2) Moreover, the traditional food-exporting countries objected strongly to what they saw as unfair competition (remember that EU products were subsidised by export refunds) in their traditional markets, and were threatening to use the forthcoming GATT negotiations to try to change things.

(3) Environmentalists were more politically powerful than they had been 20 years earlier and were pointing out that the CAP was creating problems for wildlife: hedge removal, wetland drainage, reclamation of rough grazing, high nutrient levels in watercourses, and so on. At the same time, people were increasingly interested in landscapes and wildlife for recreational purposes.

These problems might have been acceptable if it could have been demonstrated that there were benefits to consumers, or poor regions, or poor farm workers. But food prices were higher than they would have been if consumers had been free to buy from the world market, poor Mediterranean regions produced products which were supported less than those produced in the richer northern regions of the EU and, in 1992, the Commission estimated that 80% of farm spending went to 20% of the farmers, 'generally the bigger and more efficient ones'. Between 1960 and 1990, the number of people engaged in agriculture in the original six member states halved, from more than 10 million to less than 5 million, and the number of farms fell from 5.9 million in 1970 to 4.7 million in 1987[1]. Clearly, the CAP was not meeting its original objectives. There was increasing pressure to change both the policy and its objectives.

10.4 CAP reform

Several attempts to change the CAP were made from the late 1970s onwards. The imposition of milk quotas in 1984 is perhaps the one best remembered. In 1991, the Commission recognised that more radical reform was required, and proposed a package of measures, the objectives of which were said to be:

(1) To maintain the Community's position as a major agricultural producer and exporter by making its farmers more competitive on home and export markets.

(2) To bring production down to levels more in line with market demand.

(3) To focus support for farmers' incomes where it is most needed.

(4) To encourage farmers to remain on the land.

(5) To protect the environment and develop the natural potential of the countryside.

Clearly, there are some conflicts between this list and the one in the Treaty of Rome. It must be assumed that this one now takes precedence. The resulting package of reforms, often known as the MacSharry reforms, since Ray MacSharry was the Commissioner responsible for agriculture at the time, was adopted in 1992. It was based on:

- Price cuts to bring EU market prices closer to world prices
- Set-aside and extensification to reduce production
- Compensation for lower returns by area or headage payments
- Measures to improve the environment and encourage forestry

and the resulting commodity regimes are outlined in the following sections.

10.5 CAP mechanisms: (1) price and income support

The **cereals** support scheme is the most basic of all the CAP regimes, and support for all the other commodities is best understood as a variation of the arrangements for cereals. Most support schemes (or *regimes* as they are often called) comprise three parts:

(1) Arrangements for domestic price support

(2) Arrangements for control of imports

(3) Compensation payments (i.e. area or headage payments)

The cereals regime distinguishes between cereals produced within the EU and those imported from *third countries* (a term often used for countries *outside* the EU), as Fig. 10.2 illustrates.

The regime has four main features:

(1) For cereals produced within the EU, the system is designed to prevent the market price from falling much below *intervention price*. In practice, it is not quite so simple, because not all grain is suitable for intervention, intervention stores are only open between November and May in northern member states, and there are monthly increments in the intervention price to com-

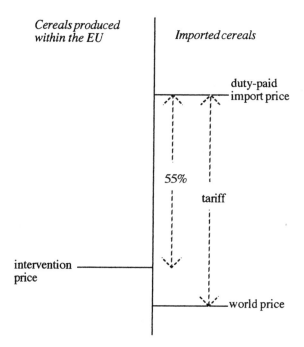

Fig. 10.2 The CAP cereals regime.

pensate for the cost of storing grain. In addition, the intervention price is denominated in ECUs, which must be converted into national currencies (see section 10.6 on green money), so the intervention price may change as often as the green rate. Grain which has been bought into intervention may be sold back on to the internal market if EU prices rise, or exported on to the world market.

(2) Traders require import licences to bring in grain from outside the EU, and when world prices are less than the *duty-paid import price* a tariff is payable. There are various complex rules, agreed in GATT negotiations, which, in theory, determine the size of the tariff. In practice, the duty-paid import price is set at 155% of the intervention price, and the *tariff* is the difference between this and the world price. Fixing the world price is the most complex part of the exercise: it involves the identification, by the Commission, of representative prices for six different categories of cereals on the first and fifteenth days of each month, with various adjustments for different qualities of grain and ports of delivery.

(3) Cereal exports are, to a greater degree than imports, controlled by the Commission, with the aim of stabilising prices within the EU. When world prices are less than EU prices, traders may tender for

export refunds; when world prices are above EU market prices the Commission uses *export levies*, calculated on the difference between domestic and world prices, to reduce the incentive to export. Again, traders may tender for export licences.

(4) Cereal producers who set aside a proportion of their land are eligible for arable area payments: in 1996, this was 274.28 ECU for each hectare of cereals grown, in England. The logic of these payments is that they are compensation for the lower prices expected as a result of the MacSharry reforms. Consequently, they differ from one region of the EU to another, to reflect the different yields expected in each region. England constitutes a single region, in Scotland the Less Favoured Areas (LFAs) are distinguished from the rest of the country, and so on; there are 90 regions in France. In order to qualify for these payments, farmers who produce more than about 15.5 hectares of cereals are required to set part (the precise proportion is agreed each year by the Agriculture Council, but is normally between 10 and 18%) of their land aside.

Thus, EU farmers are guaranteed a return for their output, and import regulations ensure that no non-EU farmers receive similar prices, while the set-aside obligation restricts the output which receives this price guarantee. The detailed regulations change from time to time, and the current regulations are contained in the explanatory documents produced by MAFF on the Integrated Administration and Control System (IACS).

Sugar is also supported by intervention, which is limited by quota, and import controls, but there is no compensation payment. *Oilseeds*, *proteins*, *linseed*, *hemp* and *flax* attract Arable Area Payments but are not bought into intervention, and neither are they subject to import controls. Imports of *fruit* and *vegetables* from third countries are subject to customs tariffs, and *withdrawal prices* are paid to producer organisations to compensate them for withdrawing surplus produce from the market and destroying it.

The basic system used for cereals also applies to *milk* and *milk products*, although it is slightly more complex because of the greater range of products involved and the impossibility of taking liquid milk into intervention. Instead, butter and skimmed milk powder (SMP) may be sold into intervention, duty-paid import prices are set for imports, and export refunds are used to assist the sale of EU–produced products on the lower-priced world market. This system, operating over many years, has resulted in significant surpluses of milk products, so national and individual *quotas* have been applied since 1984. When the national quota is exceeded, all producers who have exceeded their individual quotas are subject to a levy of 115% of the target price for their excess production. Individual quotas may be sold or leased.

The quantity of *beef* and *veal* which may be bought into intervention is limited to 350 000 tonnes, except when the market price is less than 60% of the intervention price, when safety net intervention is allowed. Imports from third countries are subject to customs duties. Headage payments are available for male cattle under the *Beef Special Premium Scheme*. An individual producer may claim for up to 90 animals in each category per year, and each animal must have a *Cattle Identification Document* (CID). The *Suckler Cow Premium* is payable to producers having suckler cow premium rights (quotas) on cows kept for producing beef calves. Both of these schemes were subject to maximum stocking densities per hectare of 2.5 livestock units in 1995 and 2.0 livestock units in 1996.

Sheepmeat production is supported by *aids to private storage* which are payable to traders if market prices fall below 85% of an annually-negotiated *basic price*. The *Sheep Annual Premium Scheme* provides for headage payments, the size of which depends on the difference between the basic price and the EU average market price, to producers who have rights to premium (quotas).

There are no headage payments for *pigmeat* producers. Pigmeat prices are protected from competition from countries where feedstuffs prices are lower by a system of *sluicegate prices* (reflecting world production costs), a *basic import levy* and a *supplementary levy* charged on pigmeat offered at less than the sluicegate price. Exports are promoted by export refunds and the internal market is supported, albeit infrequently, by aids to private storage. *Eggs* and *poultrymeat* producers are protected from third country competition by the same mechanism as that used for pigs, but there is no internal market support.

⇨ **Questions**_____

See if you can match the definitions (1–10) with the technical terms (A–J) below:

1 The price at which domestically produced cereals can be taken off the market
2 The lowest price at which foreign cereals can be sold in the EU
3 The difference between the duty-paid import price and the world price of cereals
4 The difference between the internal market price for cereals and the world price, paid on exports
5 The pig and poultry equivalent of the duty-paid import price

6 Like intervention, but paid to encourage the meat trade to keep surpluses off the market
7 The penalty for exceeding your milk quota
8 Compensation per ton of cereals multiplied by regional average yield
9 A headage payment paid only on male cattle
10 The Integrated Administration and Control System
A Arable Area Payments for cereals
B Sluicegate price
C Beef Special Premium
D Aids to private storage
E Duty-paid import price
F IACS
G Superlevy
H Export refund
I Intervention price
J Tariff for cereals

11 Use the information in the text to fill in the table below.

Enterprise	Price supported by	Imports controlled by	Compensation payments
Cereals			
Oilseeds			
Protein crops			
Linseed, hemp, flax			
Sugar			
Potatoes			
Milk and milk products			
Beef and veal			
Sheepmeat			
Pigmeat			
Fruit and vegetables			

10.6   CAP mechanisms: (2) green money

Before you can understand how green money works, you need to know something about *the theory of exchange rates*, because the green money system is really nothing more than a particular sort of exchange rate.

The easiest way to understand exchange rates is to think of them as the price of a currency. Therefore, like any other price, they are determined by the forces of supply and demand. The quantity on the horizontal axis of Fig. 10.3 is the quantity of pounds on the international money markets.

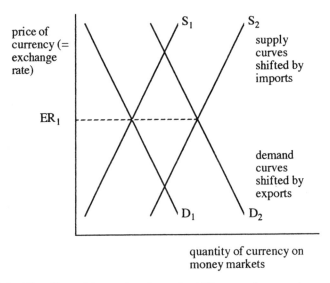

Fig. 10.3   The effect of demand and supply shifts on exchange rates.

Using Fig. 10.3, we can ask the following:

(1) *What shifts the supply curve?* Suppose I go on holiday to France and bring back a crate of wine. I will have to pay for it in French francs, so I go to the money market, supply it with pounds, and ask for francs in return. The supply curve of pounds sterling shifts to the right (S_1 to S_2). So *imports* shift the *supply* curve.

(2) *What shifts the demand curve?* Suppose a Frenchman comes to England, walks into Marks and Spencer, and buys a pullover. He will have to pay for it in pounds sterling, so he will go to the money market, demand pounds, and exchange them for francs. The demand curve for pounds shifts to the right (D_1 to D_2). So *exports* shift the *demand* curve.

Thus, the balance of imports and exports – the Balance of Payments

– affects the exchange rate (ER). If exports exceed imports (i.e. the demand curve moves further to the right than the supply curve) then the exchange rate will rise. If imports exceed exports the exchange rate will fall. Thus, a country with a Balance of Payments deficit will find its exchange rate falling if everything else stays constant. In practice, other things rarely stay constant. The demand and supply curves can also be affected by capital movements (themselves affected by interest rates) and expectations of future rates of inflation.

⇨ **Questions**_____

Now complete the following situations:

12 If the UK becomes less self-sufficient in food products, imports will *rise/fall* and the *demand curve/supply curve* for pounds sterling on the international money markets will move to the *left/right*, so that the exchange rate of the pound will *rise/fall*.

13 If interest rates in the UK rise, foreign investors will find it more desirable to put their money into the UK, so the *demand curve/supply curve* for pounds sterling on the international money markets will move to the *left/right*, so that the exchange rate of the pound will *rise/fall*.

14 If the rate of inflation in the UK rises, foreign investors will worry that the value of their holdings will decrease, and so will find it less desirable to put their money into the UK, so the *demand curve/supply curve* for pounds sterling on the international money markets will move to the *left/right*, so that the exchange rate of the pound will *rise/fall*.

If you have grasped these ideas about exchange rates, you should now be able to see that, unless there is some mechanism to prevent them from doing so (such as Economic and Monetary Union), they will vary slightly all the time. This has several implications, but the most important one, as far as agriculture is concerned, is that prices denominated in European Currency Units (ECUs) would vary constantly if ECUs were exchanged into national currencies at money market exchange rates. Imagine the chaos that would result at intervention stores! Consequently, the EU fixes exchange rates to be used for agricultural purposes. These are the **green rates**, which are fixed for a period of time. When they change, the price received by farmers changes too. Imagine a commodity with an intervention price of 100 ECUs per tonne:

Intervention price (ECUs)	×	Green rate (1 ECU =)	=	Intervention price (£ sterling)
100		0.84		84

⇨ **Question**

15 What would the sterling intervention price be if (a) the green rate devalued to 0.88, and (b) the green rate revalued to 0.82?

Another problem can arise when green rates of exchange differ from market rates of exchange. If the difference were sufficiently great, it could become worthwhile for farmers to sell into intervention in a country other than their own, as the following (completely fictitious) example of a product which sells into intervention for 100 ECU per tonne demonstrates:

	France (FF)	UK (£)
Market rate (1 ECU =)	8	0.88
Green rate (1 ECU =)	8	0.84
Sell into intervention in own country for	800	84.00

But if the UK farmer sells into intervention in France he receives:

800 FF worth £ [800 × (0.88/8)] = £88

So as long as the transport costs to France were less than £4 per tonne, it would be worthwhile to sell to a French intervention store. The EU does not wish to encourage such purely artificial trade flows and therefore has a mechanism to prevent the gap between the representative market rate (which is the daily market exchange rate for each currency, averaged over a 10 day reference period) and the green rate – known as the *real monetary gap* (RMG) – from becoming too large. The real monetary gap is calculated using the expression:

$$RMG = \frac{\text{Green rate} - \text{Representative market rate}}{\text{Green rate}} \times 100$$

Thus, if the green rate is 1 ECU = £0.84 and the representative market rate (RMR) is 1 ECU = £0.88, the real monetary gap will be:

$$\frac{0.84 - 0.88}{0.84} \times 100 = -4.76190$$

Since the real monetary gaps are not allowed to exceed +5% or –2% (with a maximum spread of gaps of 5 percentage points), the green rate would have to be halved, using the formula

$$\text{RMR} \times \frac{100}{(100 - \text{RMG}/2)} = \text{New green rate}$$

$$\text{i.e. } 0.88 \times \frac{100}{[100 - (-4.76190/2)]} = 0.85953$$

If the resultant gap is still outside the boundary value of –2% it has to be halved again, until it *is* less.

⇨ **Question** _____

16 Is the monetary gap still outside the boundary value of –2%? If it is, and it has to be halved again, what would the new green rate be?

These green rate changes are made at the end of each reference period, in conformation to rules which aim to maintain the RMGs at between +5% and –2%, with a maximum spread of 5 percentage points (i.e. if the maximum positive gap were +5%, negative gaps would not be permitted).

Since green rate revaluations to take account of positive real monetary gaps have the effect of reducing intervention prices and compensation payments, they are politically more sensitive than devaluations of negative gaps. In consequence, some revaluations can be delayed for up to five reference periods. The arrangements are explained in detail in the *CAP Monitor* (Agra Europe, London, see Chapter 11).

The December 1995 meeting of the European Council in Madrid confirmed 1 January 1999 as the starting date for stage 3 of Economic and Monetary Union (EMU), so from then onwards CAP institutional prices should be expressed in Euro. For EMU participants, exchange rates against the Euro will be fixed and so separate green rates will be redundant. For other currencies the situation is more complex, but it has already been decided that, until 1 January 1999, area and headage payments will be paid at the green rates in force on 23 June 1995.

10.7 CAP mechanisms: (3) structural and environmental policy

'Structural' in this case means 'concerned with farm size structure' and the things associated with it, such as investment, marketing and problem agricultural areas.

For historical reasons, there are many small farm businesses in the EU. A small business is not necessarily less efficient than a big one, but it will have more difficulty in surviving at any given product price level because it produces less units of output to cover its fixed costs. The commodity policies of the CAP are not designed to solve this problem so, in 1968, the Commission Memorandum on the Reform of Agriculture – known, after the then Commissioner for Agriculture, as the 'Mansholt Plan' – suggested that it would be easier to operate the commodity policies if there were fewer small farms, or that those which remained were more highly capitalised. Accordingly, in 1972, schemes to promote early retirement among small farmers, greater investment by remaining farmers, and the provision of advice to farmers, were introduced. In 1975, the special problems of farming in mountain, hill and Less Favoured Areas (LFAs) were recognised and, more recently, the environmental effects of farming have also been taken into account.

The LFA directive, which recognised the desirability of assisting specific regions of the EU, was followed by further measures to assist Mediterranean regions, and also by small 'integrated programmes' which, for the first time, introduced the concept of integrating agricultural and non-agricultural development in problem rural regions. This integrated approach was the one to be emphasised in 1988 when it was agreed that structural spending, through the European Regional Development Fund, the European Social Fund and the agricultural budget, should be increased. Five priority objectives were identified, the most significant of which for rural purposes were Objective 1, for the development of regions whose per capita GDP was less than 75% of the EU average, and Objective 5b, for low-income agricultural areas (which include parts of Devon and Cornwall, mid-Wales and western Scotland). In these areas a series of Operational Programmes have been established which promote not only agricultural diversification but also tourism, environmental protection and training.

All previous structural policy measures have now been replaced by Regulation 2328/91, which provides funds for co-ops and producer groups, technical training, improvements in marketing and processing, investment grants and compensatory payments for farmers in LFAs. In the UK, most structural funds are channelled through the Farm and Conservation Grant Scheme and funds for Objective 5b areas.

Environmental measures are carried out under the provisions of Regulation 2078/92 (sometimes known as the Agri-Environment Regulation). In the UK, it is applied through the Countryside Stewardship, Environmentally Sensitive Areas, Nitrate Sensitive Areas, Countryside Access, Organic Aid, Moorland, and Habitat Schemes. Regulation 2080/92 on afforestation on farms is applied through the Woodland Grant Scheme of the Forestry Authority.

10.8 The decision-making mechanisms of the EU

It should be clear, from the foregoing description of the CAP, that many decisions must be made constantly for it to operate successfully. Consequently, if we want to understand why the CAP works in the way it does, it is important to have a basic idea of the decision-making mechanisms by which the EU takes into account the wishes of more than 370 million people, their lobbies and pressure groups.

These mechanisms are summarised in Fig. 10.4. It shows how the various institutions of the community combine together to propose and decide upon primary legislation.

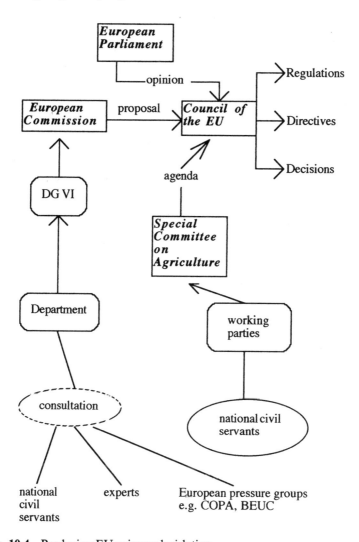

Fig. 10.4 Producing EU primary legislation.

The institutions of the EU

The four main institutions involved are:

(1) *The European Commission* is presided over by 20 Commissioners, appointed by the governments of the member states but required to act in the interests of the EU as a whole. Each Commissioner is responsible for a specific area of policy (e.g. there is one for agriculture and another for environmental issues). The EU's civil servants also work for the Commission and are organised into Directorates-General (DGs). *DG VI* (Directorate-General No.6) is the DG in which the civil servants responsible for the administration of the CAP work.

(2) *The Council of the European Union* is a meeting of the departmental ministers responsible for a specific area of policy. Thus, agricultural policy is decided in the agriculture council which consists of the Ministers of Agriculture of the member states. It meets once a month, usually in Brussels.

(3) *The SCA* is the Special Committee on Agriculture, which is a meeting of national civil servants from each of the member states whose job is to sort out which of the proposals produced by the Commission are controversial.

(4) *The European Parliament* is directly elected by the people of the EU. The members are grouped into several parties, the Socialists, Social Democrats and Christian Democrats being among the biggest.

The primary legislation process

The legislative process begins with the Commission. In agricultural matters, initial proposals are produced by the appropriate department of *DG VI*, usually in consultation with national civil servants, independent experts, and a wide variety of pressure groups such as *COPA* (the Comite des Organisations Professionelles Agricoles, i.e. the association of European farmers' pressure groups) and *BEUC* (which is the consumers' equivalent of COPA). The draft proposal is cleared with other DGs in the Commission and submitted to a meeting of the 20 Commissioners for their approval. Once approved, the draft becomes an official Commission proposal and is passed to the Council. Before it is discussed in a full meeting of the Council it will have been considered by the SCA: only those issues which cannot be agreed in the SCA need to be discussed by the Council. At the same time, the Parliament is asked to produce a formal Opinion on the proposal, which it does by nominating a committee to produce a report. This report may be discussed by a plenary meeting of the Parliament before being adopted as the formal Opinion. The Opinion must be received by the Council before it can make a decision. While this consultation is taking place, national pressure groups will lobby their own national government ministers who are members of the Council. Finally, if necessary, the proposal will be discussed in the Council and, if there is no unanimity, voted upon by qualified majority (which means that the bigger countries in the EU have more votes than the smaller countries). The Council's decisions must, however, be approved by the Parliament.

Three kinds of legislation emerge from the Council's deliberations. A Regulation applies to all people and governments in the EU; a Directive is binding as to its intention on governments; and a Decision is similar to a Regulation but only for those to whom it is addressed.

⇨ **Questions**_____

Match the following functions (17–21) with the EU institutions (A–D) listed below:

17   An institution which produces opinions on legislative proposals
18   The body which initiates proposals
19   The body which produces the agenda for a meeting of the Council
20   The institution mainly consulted by European pressure groups
21   The body which decides whether or not to accept the Commission's proposals

A   The Council of the European Union
B   The Special Committee on Agriculture
C   The European Parliament
D   The European Commission

10.9   The future of the CAP

The problems of the CAP before the MacSharry reforms have already been outlined earlier in this chapter, and at the time of writing it is still too soon after those reforms to produce a critique of them which is based on outcomes rather than predictions. Nevertheless, some points can be made about the likelihood of their accomplishing what they set out to do, and on the factors affecting the future development of the CAP.

The MacSharry objectives of increasing competitiveness and bringing production into line with market demand, especially in the cereals market, were to be brought about by imposing lower prices and set aside. Ironically, by the time the reformed policy was fully in place, world prices had risen above EU prices for the first time in 20 years and so European producers were indeed competitive, to the point where export levies had to be introduced to keep EU prices down. On the other hand, the benefits of this were obviously felt most by those who had most to sell, so that the position of the marginal farmers, insofar as they tended to be the small farmers, was unlikely to be improved. They were more dependent on the compensation payments but these were allocated on a headage or hectarage basis, so that, again, the bulk of the expenditure tended to go to the bigger producers. Moreover, high world cereal prices were not always replicated in other commodities. But at least the environmental objectives of the reform package were not hindered by the adoption of the various schemes introduced under the terms of the Agri-Environment Regulation.

By late 1995, the Commission had implicitly accepted that further evolution of the reformed CAP would be needed to cope with the development of the EU as a whole. The prospect of economic and monetary union, the increasing importance of environmental and animal rights issues and the planned enlargement of the EU through the accession of the Central and Eastern European Countries (CEECs) all had implications for the CAP. The CEECs, for example, have large agricultural industries, so their accession would be expected to result in significant extra agricultural spending. However, the budget guidelines allow agricultural spending to increase by only 0.74% for each 1% increase in GDP. The entry of the CEECs would only produce a relatively small increase in the size of the EU economy so that an unchanged CAP would soon be expected to run short of money. In addition to the internal changes, the CAP would also need to respond to the next round of GATT talks, planned to begin towards the end of the century.

The development of world prices is therefore crucial to the future of the CAP. High prices would imply little expenditure on export subsidies and would keep the intervention stores empty. It would be easier to comply with the EU's Uruguay Round obligations. The costs of CEEC accession would be reduced. Conversely, low world prices would increase the pressure on the EU budget. On the other hand, environmentalists would perhaps argue that high prices, if maintained for very long, would increase the pressure on wildlife habitats as farmers were tempted to increase output by reclamation and intensification.

Commentators have differed in their prediction of the outcome. Some have emphasised the influence on demand of increasing incomes and populations in the newly-industrialising countries; others have argued that the return of set-aside land to cultivation, coupled with technical changes, could rapidly increase supplies. The possibilities are constantly under discussion in the technical press: using the methods of analysis with which you became familiar in the earlier chapters of this book might enable you to decide which of the various arguments makes most sense at any particular time.

Note

1. Commission of the European Communities (1993) *Our Farming Future*, The Commission, Brussels.

Chapter 11
Epilogue

The previous chapter outlined the basic principles of the CAP, but you will probably have realised, if you have been reading the farming press, that there are many important details of policy which it does not cover. This is *not* simply because this is a short book which lacks the space to go into detail. The important thing to recognise about the CAP is that its basic principles change only slowly although the details can differ dramatically overnight as a result of one meeting of the Agriculture Coucil. Therefore, it is important to learn how the basic principles work. Once you know that, it's easy to fit the detailed changes into the fundamental framework and so keep up to date; if you are hazy about the basics it's difficult to see where the constant flow of small changes fits into the whole. Thus, the purpose of this book, and especially of Chapter 10, is to concentrate on the basics so that you have the framework to support what you learn from your reading of the press and any other sources of information you use.

If you want more detail on the CAP, or you want to know how to keep up to date with the detailed changes, look in the suggestions for further reading which form the second half of this chapter.

You may also have noticed that this book concentrates on the agricultural industry to the exclusion, in the main, of much discussion of rural society as a whole, or other land uses, or other parts of the food chain. This *is* for reasons of space. Nevertheless, you will probably have realised, if you have read the section on decision making in the EU in Chapter 10, or the section on politics in Chapter 9, that many people and pressure groups have a say in decisions which affect farmers. This is inevitable, because:

- Farmers, farm workers and landowners live in rural societies which also include other people who do not have such close ties with farming but are often affected by agricultural activities.
- Agriculture uses land which could also be used for recreation, forestry, water gathering, wildlife habitats, building houses and

factories, and military training, and sometimes several of these at once.

- Farming buys lots of inputs from the agricultural engineering, chemical and building industries (which are *upstream* to agriculture) and sells its output to the food processing and retailing industries (the *downstream* industries).

Therefore, lots of people are affected by the activities and wellbeing of the agricultural industry, and agriculture is influenced by changes in rural society, changing demands by other land-using industries, and changes in the upstream and downstream industries. Most of these changes are well known, but it is worthwhile to review some of them briefly and to asses their impact on agriculture.

The main change in UK rural society has been the reversal, since the 1960s, of the trend to rural depopulation. Most rural areas in lowland England are within commuting range of major centres of population and employment, and so many people who now live in villages no longer work in the traditional rural industries. This also applies to some parts of highland England and, where it does not, the attraction of second homes maintains the demand for rural property. In some parts of North Cornwall, for example, almost half the houses are said to be second homes. One effect of all this is to reduce the importance of agriculture in the rural economy. It can no longer be argued that rural jobs are dependent on the profitability of agriculture if the majority of those living in rural areas are working in towns. In some places it has also resulted in greater difficulties for farmers with intensive livestock enterprises: somebody who has invested their life-savings in a bijou thatched cottage with roses round the door is unlikely to be enthusiastic about a proposal to establish a 500-sow indoor pig unit in the field next door. This is one of the reasons why such units are now subject to planning controls. The continued attraction of country living has also meant that farmers have a range of objectives, from those who graze a few bullocks or sheep to keep their horses company, while they get the bulk of their income from elsewhere, to those who are concerned with maximising the return on capital invested in a large commercial operation. And this is just the variety to be found in the UK. But the CAP applies to the whole of the EU, from northern Finland to Sicily and from Portugal to Prussia. The range of rural societies to be found among the 370 million people who live there is enormous. Just because commuters have moved back to the UK countryside, it does not necessarily follow that the same pressures will keep people in the rolling hills of south-west France or the mountains of Greece. Consequently, the objectives of those who attempt to influence the agricultural policy makers will be equally varied.

One of the more noticeable trends in the last quarter of the twentieth century has been the increased interest in wildlife and landscape conservation. This is illustrated by the figures for membership of voluntary environmental organisations shown in Table 11.1.

Table 11.1   Membership of some UK voluntary environmental organisations (thousands).

Organisation	1971	1993
National Trust	278	2189
CPRE	21	45
Ramblers' Association	22	94
Woodland Trust	—	150
RSPB	98	850
RSNC	64	248
World Wide Fund for Nature	12	207
Greenpeace	—	410
Friends of the Earth	1	120

Source: CSO (1995) *Social Trends*, vol. 25, p. 189, HMSO, London.

The figures in Table 11.1 need to be handled with care because some people may be members of more than one organisation so that an element of double counting is involved. Conversely, some people may support the aims of these organisations without actually joining them. From the viewpoint of agricultural policy, the important point is that the increased membership of these organisations illustrates the rising awareness of environmental issues in the UK so when the result of agricultural policy is to produce surpluses, many people question its desirability. Similarly, when one of the aims of agricultural policy was to increase national self sufficiency, the planning mechanism was very protective of agricultural land; during the 1990s, there have been several central government initiatives in the opposite direction.

As the food retailing sector has become increasingly concentrated, with a few large firms now being responsible for a high proportion of total grocery sales, the market power of retailers and food processors has increased. The difference between an armed terrorist and a supermarket buyer, one exasperated farmer is reputed to have said, is that you can negotiate with an armed terrorist! The food processing and retailing industries are now significantly bigger than agriculture, as Table 11.2 indicates, which is perhaps not surprising when you remember that as people get richer they do not necessarily eat a greater quantity of food, but they do eat food with more value added. Clearly,

Table 11.2   Agriculture and the rest of the food chain.

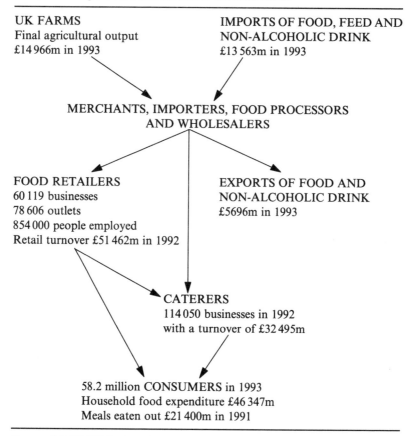

UK FARMS
Final agricultural output
£14 966m in 1993

IMPORTS OF FOOD, FEED AND
NON-ALCOHOLIC DRINK
£13 563m in 1993

MERCHANTS, IMPORTERS, FOOD PROCESSORS
AND WHOLESALERS

FOOD RETAILERS
60 119 businesses
78 606 outlets
854 000 people employed
Retail turnover £51 462m in 1992

EXPORTS OF FOOD AND
NON-ALCOHOLIC DRINK
£5696m in 1993

CATERERS
114 050 businesses in 1992
with a turnover of £32 495m

58.2 million CONSUMERS in 1993
Household food expenditure £46 347m
Meals eaten out £21 400m in 1991

Sources: MAFF (1994) *Agriculture in the UK, 1994*, Table 1.1, HMSO, London; CSO (1995) *Anual Abstract of Statistics, 1995*, Tables 2.1, 9.1, 11.1 and 11.4, HMSO, London. I am most grateful to Derek Shepherd for providing this material.

the objectives of the food industry are not necessarily the same as those of farmers: the farmer's high output price is the food processor's high input price. On the other hand, the upstream ancillary industries will sell more to a profitable expanding agriculture than to a contracting industry. Thus, the political power of the manufacturing industry may deliver conflicting messages through its lobbyists.

It should also be remembered, when examining Table 11.2, that the boundary between the food industry and agriculture is increasingly fluid. There are now large meat and egg producers who run their own production operations and farmer co-operatives which process their own vegetables.

In this book, agriculture has been characterised as a declining industry with agricultural policy a response to the excess of supply over demand. This has been the long-term trend in industrialised countries since the last two decades of the nineteenth century. Whether it will continue to be so is now beginning to be questioned. In the mid-1990s, weather-induced supply shortages have combined with demand increases produced by economic growth. The result has been an increase of world cereal prices above EU intervention price levels for the first time in 20 years: short-term fluctuation, or the beginning of a long-term trend? Opinions were divided at the time of writing. In either case, the implications for EU agriculture are profound. Rising prices would produce very different outcomes and problems from falling prices. And in this constant succession of new issues in changing circumstances lies some of the fascination of agricultural economics.

How to find out more

The issues to which you have been introduced in this book are covered in more detail in several more advanced textbooks of agricultural economics. A reference list is at the end of this chapter. The most recent of these is Michael Tracy's *Food and Agriculture in a Market Economy: An Introduction to Theory, Practice and Policy*. Rather older, but in many ways still the most thorough and readable treatment of the subject, is Berkeley Hill and Derek Ray's *Economics for Agriculture: Food, Farming and the Rural Economy*. A different approach is offered in A.J. Rayner and D. Colman's (eds) *Current Issues in Agricultural Economics* and C. Ritson and D. Harvey's (eds) *The CAP and the World Economy: Essays in Honour of John Ashton*. The main academic journals for the subject, which contain not only the results of research but also reviews of current books, are the *Journal of Agricultural Economics* and the *European Review of Agricultural Economics*. Some of the more important papers which have appeared in these and other journals in recent years have been reprinted in G.H. Peters' (ed.) *Agricultural Economics (International Library of Critical Writings in Economics: 55)*.

The more detailed and deeper approach offered in these works requires some knowledge of economic theory. There are many textbooks of economics: one of the best of the recently published is John Sloman's *Economics*.

For more information on the CAP, Rosemary Fennell's *The CAP of the EC* gives the background to the development of detailed policy measures but is, inevitably, somewhat dated. The *CAP Monitor*, a continuously updated work published by Agra Europe (London) Ltd sets out current regulations.

This book approaches the problems of the agricultural industry from an economic and historical perspective. There are several interesting works which look at the same issues from a political angle, such as Michael Winter's *Rural Politics: Policies For Agriculture, Forestry and the Environment* and H.W. Moyer and T.E. Josling's *Agricultural Policy Reform: Politics and Process in the EC and USA*. From a purely American viewpoint, there is W.P. Browne *et al.*'s *Sacred Cows and Hot Potatoes: Agrarian Myths in Agricultural Policy* and C. Hamlin and P.T. Shephard's *Deep Disagreement in US Agriculture: Making Sense of Policy Conflict*. A sociological approach also has some useful insights to offer. Howard Newby's *Green and Pleasant Land?* sets the problems of agriculture in the context of the countryside in general. Ronald Blythe's classic *Akenfield* offers a detailed portrait of an individual rural community, as does Michael Mayerfeld Bell's *Childerley: Nature and Morality in a Country Village* which also surveys some of the more recent work on rural sociology. K. Hoggart, H. Buller and R. Black's *Rural Europe* is largely written from a geographical angle.

The historical development of agriculture in the EU, over the last two centuries at least, is dealt with in Michael Tracy's *Government and Agriculture in Western Europe*. The agricultural history of England and Wales is covered in Joan Thirsk's (ed.) *The Agrarian History of England and Wales* (eight volumes, of which vol. vii has yet to appear). Dr Thirsk takes an interestingly different approach in *Alternative Agriculture, Past and Present*. In a briefer compass, recent history is covered in David Grigg's *English Agriculture: An Historical Perspective* and Quentin Seddon's *The Silent Revolution*, which mainly covers the period since World War II and is journalism rather than history, but good journalism. There are several interesting biographical approaches to recent agricultural history, one of the best of which is John Cherrington's *On the Smell of an Oily Rag*.

You can bring the statistics in this book up to date from the Ministry of Agriculture's *Agriculture in the UK, Farm Incomes in the UK*, and *A Digest of UK Agricultural Statistics* (all published annually). The EU statistics office, Eurostat, publishes *Agriculture: Statistical Yearbook* annually, and there is also an annual survey of *The Agricultural Situation in the Community*. John Nix's annual *Farm Management Pocketbook* contains a brief but useful statistical section and is, of course, invaluable for data on production costs and details of area and headage payments. Longer runs of statistics may be found in A. Burrell, B. Hill and J. Medland's *Agrifacts*, H.F. Marks and D.K. Britton's *A Hundred Years of British Food and Farming: A Statistical Survey*, H.F. Marks' *Food: Its Production, Marketing and Consumption*, and MAFF *A Century of Agricultural Statistics* (this was published to celebrate the centenary of the annual agricultural census in the UK and contains material which is still difficult to find elsewhere). There are two relevant volumes in the series of *Reviews of UK Statistical Sources:* vol. xxiii, *Agriculture*, by G.H. Peters and vol. xxviii, *The Food Industries* by J. Marks and R. Strange, both of which not only tell you where to find data but alert you to some of the problems of using it.

Finally, the best source of up-to-date news and data on the EU is the weekly *Agra Europe*, which is expensive but unrivalled and often the source used by the rest of the agricultural press. The Meat and Livestock Commission and the Home Grown Cereals Authority produce various weekly and monthly statistical publications on the commodities with which they are concerned. If you are a member of the human race at 6.10 a.m., the BBC's *Farming Today* programme on Radio 4 is perhaps the most up-to-date source of information on major policy developments.

References to Chapter 11

W.P. Browne *et al.* (1992) *Sacred Cows and Hot Potatoes: Agrarian Myths in Agricultural Policy*, Westview Press, Oxford.

A. Burrell, B. Hill and J. Medland (1990) *Agrifacts*, Harvester Wheatsheaf, Hemel Hempstead.

J. Cherrington (1979) *On the Smell of an Oily Rag*, Northwood, London.

Eurostat (Annual) *Agriculture: Statistical Yearbook*, Brussels.

Eurostat (Annual) *The Agricultural Situation in the Community*, Brussels.

R. Fennell (1987) *The CAP of the EC*, 2nd edn., Blackwell, Oxford.

D. Grigg (1989) *English Agriculture: An Historical Perspective*, Blackwell, Oxford.

C. Hamlin and P.T. Shephard (1993) *Deep Disagreement in US Agriculture: Making Sense of Policy Conflict*, Westview Press, Oxford.

B. Hill and D. Ray (1987) *Economics for Agriculture: Food, Farming and the Rural Economy*, Macmillan, London.

K. Hoggart, H. Buller and R. Black (1995) *Rural Europe*, Arnold, London.

MAFF (1967) *A Century of Agricultural Statistics*, HMSO, London.

MAFF (Annual) *A Digest of UK Agricultural Statistics*, HMSO, London.

MAFF (Annual) *Agriculture in the UK*, HMSO, London.

MAFF (Annual) *Farm Incomes in the UK*, HMSO, London.

J. Marks and R. Strange (1993) *The Food Industries*, vol. xxviii, Reviews of UK Statistical Sources, Chapman and Hall, London.

H.F. Marks (1992) *Food: Its Production, Marketing and Consumption*, Taylor and Francis, London.

H.F. Marks and D.K. Britton (1989) *A Hundred Years of British Food and Farming: A Statistical Survey*, Taylor and Francis, London.

M. Mayerfeld Bell (1994) *Childerley: Nature and Morality in a Country Village*, University of Chicago Press, Illinois.

H.W. Moyer and T.E. Josling (1990) *Agricultural Policy Reform: Politics and Process in the EC and USA*, Harvester Wheatsheaf, Hemel Hempstead.

H. Newby (1979) *Green and Pleasant Land?*, Penguin, Harmondsworth.

J. Nix (1996) *Farm Management Pocketbook*, Wye College Press, Ashford.

G.H. Peters (1988) *Agriculture*, vol. xxiii, Reviews of UK Statistical Sources, Chapman and Hall, London.

G.H. Peters (ed.) (1995) *Agricultural Economics (International Library of Critical Writings in Economics: 55)*, Edward Elgar Publishing, Cheltenham.

A.J. Rayner and D. Colman (eds) (1993) *Current Issues in Agricultural Economics*, Macmillan, London.

C. Ritson and D. Harvey (eds) (1991) *The CAP and the World Economy: Essays in Honour of John Ashton*, CAB, Wallingford.

Q. Seddon (1989) *The Silent Revolution*, BBC Books, London.

J. Sloman (1994) *Economics*, 2nd edn., Prentice-Hall, Englewood Cliffs.

The CAP Monitor, Agra Europe (London) Ltd., London.

J. Thirsk (ed.) (since 1967) *The Agrarian History of England and Wales*, Cambridge University Press, Cambridge.

J. Thirsk (In press) *Alternative Agriculture, Past, Present and Future*, Oxford University Press, Oxford.

M. Tracy (1989) *Government and Agriculture in Western Europe*, 3rd edn., Harvester Wheatsheaf, Hemel Hempstead.

M. Tracy (1993) *Food and Agriculture in a Market Economy: An Introduction to Theory, Practice and Policy*, Agricultural Policy Studies, La Hutte, Belgium.

M. Winter (1996) *Rural Politics: Policies for Agriculture, Forestry and the Environment*, Routledge, London.

ANSWERS TO QUESTIONS

Chapter 2: How markets work

1 Your graph should look like this:

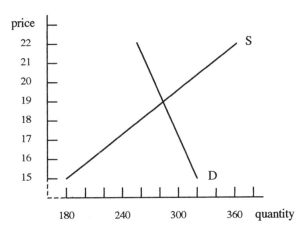

S = supply, D = demand.
At 15 ppl, $Q_d = 320$ and $Q_s = 180$, so the excess Q_d is $(320 - 180 =)$ 140 million litres, or, in other words, there is a shortage of 140 million litres per week at 15 ppl.

2 At 22 ppl, $Q_d = 256$ and $Q_s = 364$, so there is a surplus of $(364 - 256 =)$ 108 million litres.

3 The price at which $Q_d = Q_s$ is 19 ppl, and the equilibrium quantity is 285 million litres.

4 Your new graph should look like this, with D_2 as the new demand curve created by the 40 million litre increase in demand.

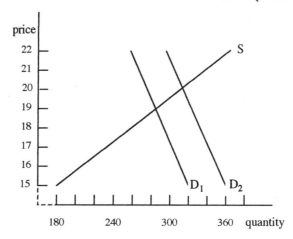

So the new equilibrium price will be 20 ppl.

5 The decline in the popularity of cooked breakfasts is a change in *tastes*, which affects demand for fat pigs.

6 An increase in the price of chicken is a change in the *price of a substitute* for the product, which is pigmeat, and this will affect demand. In the past, it might have been possible to argue that it was also a change in the price of other products which the farm might produce, which would affect supply, but the production of poultrymeat is now such a specialist business that pig producers are unlikely to respond to an increase in chicken prices by moving out of pig production in order to produce broilers.

7 An increase in the price of fishmeal is an increase in the costs of production, which affects supply, since fishmeal is a source of protein in concentrates.

8 When the price falls to P_2, the quantity supplied moves down the curve to Q_{s2}, and the quantity demanded increases to Q_{d2}.

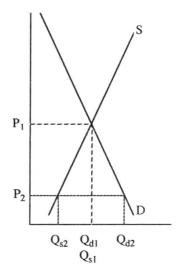

Fig. 2.2 redrawn to illustrate what happens if the price falls.

9   A decrease in income shifts the demand curve from D_1 to D_2.

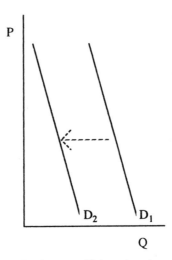

Fig. 2.3 redrawn to illustrate what happens if there is a decrease in income.

10   Output-increasing technology shifts the supply curve from S_1 to S_2, so the curves now intersect at Q_{d2} and Q_{s2}. The price falls to P_2.

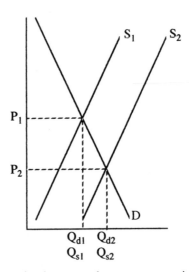

Fig. 2.2 redrawn to illustrate what happens when new, output-increasing technology is introduced.

The table below shows the answers to questions (11) to (20).

Question	Factor affected	Affects demand (D) or supply (S)	Movement along (MA) or shift (Sh) in the curve	Shift left (L) or right (R)	Changes in equilibrium price (P) and quantity (Q)
11	Consumer income	D	Sh	R	P up, Q up
12	Population	D	Sh	R	P up, Q up
13	Price of alternative products	S	Sh (but not much)	L	P up, Q down
14	Costs of production	S	Sh	L	P up, Q down
15	Price of product	D, S	MA	—	P up, Q_d down, Q_s up
16	Technology	S	Sh	R	P down, Q up
17	Price of substitutes	D	Sh	L*	P down, Q down
18	Tastes	D	Sh	L	P down, Q down
19	Unpredictable	S	Sh	L	P up, Q down
20	Price of product	D, S	MA	—	P down, Q_d up, Q_s down

* The explanation for question 17 is that as margarine prices decrease, so the quantity of margarine bought increases, and the quantity of butter purchased at any price decreases, thus shifting the butter demand curve to the left.

21 The price elasticity of demand is higher for steak than for meat products in general because there are more substitutes for steak.

22 The cross-price elasticity of demand for complementary products is negative because, as the price of beer rises, the quantity demanded falls, so presumably the quantity of crisps demanded will also fall. Thus, a price increase has brought about a quantity decrease, and so the ratio in the elasticity calculation will be negative.

23 The income elasticity of demand will be higher for gateau, which is a luxury product.

24 It will take longer for the supply of timber to respond to a price change, simply because it takes much longer to produce timber than wheat, so the price elasticity of supply will be higher for wheat.

25 The change in quantity is –8%, and the change in price is +5%, so, using the formula in the handout, the price elasticity of demand will be $-8/+5 = -1.6$.

26 $-1.5 = Q_d/20$, so $Q_d = -1.5 \times 20 = 30\%$.

27 Income elasticity = 0.2, change in income = 3%, so change in quantity = $3 \times 0.2 = 0.6$.

28 Elasticity of supply = $1.5/5 = 0.3$.

29 If the price increased by 1%, the quantity would decrease by less than 1%, so the increase in price would be more than the decrease in quantity, and the total revenue would therefore increase. We shall look at another way of demonstrating this important result in the next chapter.

Don't worry if you got some of these wrong at the first attempt. There are some testing questions here. Just make sure that you know why you went wrong. Once you can see how to do most of these questions you will have a pretty good grasp of the main theoretical content of the whole book.

Chapter 3: The demand for agricultural products

1   From Table 3.1, the price elasticity of demand for bread is –0.09, so a 10% increase in the price of bread would decrease the quantity demanded by (–0.09 × +10 =) –0.9%.

2   The price elasticity of demand for liquid milk is –0.29, so the price decrease required is (+10/–0.29 =) 34.48%.

3   There are more substitutes for UK butter (such as Danish or New Zealand butter) than there are for butter as a whole, so if the price of UK butter goes up consumers might buy these substitutes instead, whereas if the prices of all butters rise, they may just keep on buying roughly the same amount.

4   Because the price elasticity of demand for potatoes is low.

5   It would have increased: quantity decreased by only 20% but price rose by 300%.

6   If the price of lamb rises, the quantity demanded will fall, so the quantity of mint sauce demanded will fall too, so the cross-price elasticity will be negative.

7   An increase in the butter price will decrease the quantity demanded, so people will turn to margarine instead and the demand curve for margarine will shift to the right.

8   The income elasticity of demand for apples is 0.32, so a 3% increase in income will decrease the demand for apples by 0.32 × 3 = 0.96%.

9   Income rises, so the demand curve for beef, which has a positive income elasticity of demand, shifts to the right, but not by much – only 0.08% for a 1% rise in incomes.

10 Income is higher in the UK so the income elasticity of demand for bread will be higher in India.

11 If the campaign is successful, the demand curve will shift to the right and prices will rise.

12 (a) Stainless steel bulk tanks are not consistent with the images advertisers usually try to associate with food products. Often, they try to find associations with health and traditional farming methods.

    (b) Contented cows in flower-strewn meadows give positive animal welfare and conservation associations, or am I just being cynical?

13 To the right, but not by much.

14 The main influences from 1950 to 1970 seem to be increasing incomes – as people got richer they ate more – and health worries decreasing demand for some products between 1970 and 1990.

Chapter 4: Changes in agricultural output

1 The first thing to do is to choose the base year. The question instructed you to choose 1985, but you might have used any other year for which you had the necessary data.

Then you have to tabulate the data and work out the indices. The results of doing this are shown in the table below:

Wheat Production

Year	France '000s tons	France Index (1985 = 100)	Germany '000s tons	Germany Index (1985 = 100)
1905	9 110	31.55	4 187	42.44
1938	9 800	33.94	6 250	63.35
1960	11 010	38.14	6 421	65.08
1970	12 921	44.75	7 794	79.00
1985	28 871	100.00	9 866	100.00
1992	30 613	106.03	15 472	156.82

Thus, overall, it is clear that wheat production increased more in Germany than in France between 1905 and 1992. But the indices also demonstrate that the story is more complex than that. Between 1905 and 1938, French production increased very little, whereas German production increased a lot, presumably as a result, in part at least, of the Nazi drive for self-sufficiency. Then both economies, recovering from World War II, experienced only small increases up to 1960. Both expanded production in the 1960s but the big increase in France came in the period 1970–85, whereas the Germans experienced much more rapid growth after 1985.

You might have seen this by careful inspection of the tonnage figures, but it is much more obvious when they have been converted into indices.

2 First tabulate the data, using the RPIs from Table 4.8.

Cost per day's hire (£)			
From Blue Star Garages in 1957		From Avis Car Rentals in 1994	
Morris Minor	1.375	Rover Metro	41.0
Austin Cambridge	1.375	Rover 416	57.0
RPI 1957	12.6	RPI 1994	152.4

Then insert the data in the formula:

$$\text{Constant price for the year} = \text{Current price for the year} \times \frac{\text{Base year RPI} (= 100)}{\text{RPI for the year}}$$

So for a Minor in 1957 the price in 1985 £s was:

$$1.375 \times \frac{100}{12.6} = 10.91$$

Thus, perhaps surprisingly, it appears that the cost of car hire has increased, as we can see when we express the prices in constant terms, as in the table below.

Cost per day's hire in constant (1985) prices (£)			
From Blue Star Garages in 1957		From Avis Car Rentals in 1994	
Morris Minor	10.91	Rover Metro	26.90
Austin Cambridge	10.91	Rover 416	37.40

On the other hand, what the question doesn't tell you is that the terms of care hire have changed since 1957. In those days, you also paid an excess mileage fee of 3d per mile (= 1.25p in decimal coinage) for every mile in excess of 25, which, in 1985 prices, was nearly 10p per mile (strictly speaking 9.92p). So if you did (160 + 25 =) 185 miles in the day the real cost of a small car would have been the same in both years, and at anything over that the 1994 price would have been cheaper. For a large car, the corresponding figure was (270 + 25 =) 295 miles.

Chapter 5: The supply of farm products

1 The return from selling 7.25 tonnes of wheat at £90 per tonne is £652.5. With the addition of the Area Payments, the total return is £902.5. Total costs are £(565 + 220 =) £785, so the farm is covering all costs. If the Area Payments are now withdrawn, the return is £652.5, so the crop is not covering its total costs, although it is covering variable costs quite easily. Therefore it should continue production in the short term, but in the long term, if fixed costs cannot be reduced, it should cease production.

2 The quantity of pigs supplied to the market will decrease as the AAPP decreases, because some producers will decrease their production and others will cease production altogether.

3 The main fixed costs on a hill farm are labour costs, which account for about half of the total fixed costs on such farms (according to J. Nix (1994) *Farm Management Pocketbook*, 25th edn., p. 138).

4 In the short term the beef supply curve would have shifted to the right, as more cull cow beef was supplied to the market, and so the price would fall. In the long term it would shift to the left as fewer calves were available for fattening, and so the price would rise. The beef production figures for the years 1982–8 (in thousands of tonnes) appear to support this hypothesis:

1982	980
1983	1044
1984	1136
1985	1132

1986	1062
1987	1118
1988	947

5 Since land will be diverted from wheat to barley, you would expect less wheat to be produced at any given price, so the supply curve will shift to the left.

6 Fertiliser cost increases will shift the cereal supply curve to the left: less will be produced at any given price because some farmers will cut down on their fertiliser use, and a few marginal producers, who were only just finding it worthwhile to stay in grain production anyway, will find that this additional cost increase means that they can no longer make enough profit to make it attractive to produce grain any longer.

7 You would expect an increase in the cost of purchased feedingstuffs to have a greater effect on the supply of milk because they form a greater proportion of the costs of milk production than AI does. According to John Nix (1995) *Farm Management Pocketbook*, 25th edn., pp. 47, 50, concentrate costs for the average cow are £206, whereas AI and recording fees are £25.

8 You would expect a lowland arable farm to have the greater proportion of bought-in inputs than a hill livestock farm.

9 You can explain these figures by shifting the supply curve to the right (as a result of technical change) faster than the demand curve shifts to the right.

10 When the demand curve is price elastic, shifting the supply curve to the right means that the quantity sold increases by a greater percentage than the price decreases, so the total revenue increases.

Chapter 6: Agricultural inputs

1 Because there is now little demand for the product they used to make, wooden wagon wheels.

2 Insofar as pigs are not a land-using enterprise, an increase in pig prices would not have much effect on rents. If pig farmers responded to higher prices by keeping more pigs outdoors it would shift the demand curve for land to the right. But the main effect would probably come through a greater demand for feed, which, as long as it was not imported, would result in a greater demand for cereal growing land and so push rents up.

3 It would shift the supply curve for *agricultural* land to the left as former agricultural land went into forestry, and so the price of farmland would rise.

4 Because not all hectares or acres are equally productive. A hectare of grade 1 land in the Fens might produce several tons of highly-priced vegetables; a couple of hours' drive away on top of the Pennines in Derbyshire, a hectare might provide only a bit of sheep grazing and some heather for the grouse.

5 We mean that their output per hectare is high because they produce livestock such as pigs or horticultural crops such as lettuce, tomatoes and flowers.

6 Because, at a given level of efficiency (= output *divided by* input) it will produce more profit (= output *minus* input). Obviously, there are some

inefficient big farms which are less profitable than some very well-run small farms.

7 It was probably affected by the increasing pressure to do something about surpluses in the EU. The introduction of milk quotas in 1984 provided a clear signal to the land market that continuing high prices supported by unlimited intervention could not be relied upon indefinitely.

8 You would not expect the number of workers to fall as rapidly as it has in the past, because high prices (assuming they affected EU prices) would lessen the push factors, while high industrial employment resulting from the depression would reduce the pull factors.

9 Because they have to pay the cost of the borrowing out of profits, and although they may have a lot of net worth, they cannot always guarantee high profit levels over long periods. In any case, the high net worth figure is a national average. Those who most need to borrow may already have high debt levels or be tenants with no land of their own.

10 Because low farm profits often mean that farmers cut back on machinery purchases, and vice versa.

Chapter 7: International trade in agricultural products

1 This question looks so simple as to be an insult to the intelligence, especially if you have read question 2 before you begin to answer it. Quite obviously, the CAP attempts to maintain EU domestic prices by excluding cheap imports from the world market (we shall examine this in much more detail later in Chapter 10). But remember, *within* the EU, there is free trade between member states, and that one of the reasons for establishing the EU was to get the benefits of free trade, not only in agricultural products but in other areas of economic activity too.

2 The main arguments used to justify protection have been concerned with protecting declining industries, maintaining food self-sufficiency in case of war and preserving ways of life based on traditional industries. We shall discuss this in greater detail in Chapter 8.

3 It is a non-tariff barrier, because it is a simple ban on imports, rather than the imposition of import taxes.

4

	Majority of exports from	
Majority of imports to	Industrialised countries	Low income countries
Industrialised countries	Meat Wine Citrus fruit *(total value of trade in this box = $35.08bn)*	Sugar Oilseeds Coffee Cocoa Rubber Bananas Cassava *(total value of trade in this box = $18.62bn)*
Low income countries	Wheat Coarse grains Cotton *(total value of trade in this box = $41.92bn)*	Tobacco Rice Tea *(total value of trade in this box = $11.95bn)*

5 The most likely candidates are the tropical beverages – coffee, cocoa and tea – because exports of these account for a *high* proportion of total production.

Appendix

A1 The lowest opportunity cost of milk production is in Souwestia (36/60W). Thus, if East Angula produces its own milk it has to give up 40/60W (= 0.67 tonnes of wheat), but if it imports from Souwestia it can obtain milk for as little as 36/60W (= 0.60 tonnes of wheat). East Angula can therefore obtain its milk more cheaply from imports than from home production.

A2 (a) Production levels are higher in Gaul, so therefore it has the absolute advantage in the production of both products.

To answer the next three parts of the question, construct an opportunity cost table. In Gaul, 6 menhirs are equivalent to 4 wild boar, so 1 menhir is equivalent to 4/6 wild boar, and so on.

	Gaul	Britannia
Menhirs	4/6WB	2/4WB
Wild boar	6/4M	4/2M

and to make comparison easier, convert to the same denominators

	Gaul	Britannia
Menhirs	8/12WB	6/12WB
Wild boar	6/4M	8/4M

and then use this table to answer the questions.

(b) Britannia has the comparative advantage in menhir production because 6/12 is less than 8/12, so the opportunity cost of menhir production is lower in Britannia.

(c) Gaul has the comparative advantage in wild boar because 6/4 is less than 8/4, so the opportunity cost of wild boar production is lower in Gaul.

(d) Britain will be better off by importing wild boar from Gaul, paying for them by exporting menhirs.

Chapter 8: The case for supporting farm incomes

1 None, except insofar as consumers buy tropical fruit *instead* of domestically produced fruit such as apples, pears and strawberries.

2 Not really. There may be political reasons, if it is predicted that political problems might arise from dependency on imports, and this case has been advanced for a whole range of products, from military aircraft and uranium ore to food. But as far as economic reasons are concerned, comparative advantage theory suggests that there is nothing magic at all about the 100% figure. If a country has a comparative advantage in the product it will be more than 100% self-sufficient, and if it has not it will be less than 100% self-sufficient if market forces are allowed to work.

3 The Balance of Payments is the balance of a country's imports and exports. If it is in surplus, or positive, it means that exports exceed imports; if in deficit, or negative, imports exceed exports. Both imports and exports are commonly divided into 'visible' and 'invisible' trade. Visible trade includes goods which physically exist, such as agricultural products, machinery, cars, TVs, and so on. Invisibles are services, such as banking, insurance, shipping and other services. The UK normally has a surplus on invisible trade and a deficit on visible trade.

4 At world price levels, the level of agricultural output would be lower than would be the case if domestic prices were maintained above world prices, and so the self-sufficiency level would change. Using only indigenous inputs would also affect the possible output level if significant quantities of feedingstuffs and fertilisers were imported.

5 There are many reasons for fluctuations in farm incomes, but variations in the quantity of output (often due to weather conditions) and the price at which it is sold are the main ones.

6 Because income elasticities of demand for farm products are low, so demand grows more slowly than supply, and prices fall faster than inputs leave the industry, as explained in the introduction to Chapter 8 and in Fig. 1.1.

7 This is an extremely difficult question. Obviously, it is not difficult to recognise a rural area – you see fields, and trees, and not so many houses and shops as in a town. But where do you draw the boundaries? And having done so, are they compatible with the sort of statistical areas for which employ-

ment data are collected? I. Hodge and S. Monk used cluster analysis on a range of population, employment, agricultural and locational variables for 257 non-metropolitan local authority districts in England to 'attempt to investigate the main economic processes of local economies' (p. 1). Some of their results are shown in the table below:

	Percentage of employees in agriculture and agriculture-related employment		
	Agriculture	Agriculture related	Number of districts
All districts	3.3	3.6	257
Parks and countryside	6.9	5.8	18
Farming	7.4	6.6	37
Service	3.5	2.5	49

Source: I. Hodge and S. Monk (1991) *In Search of a Rural Economy: Patterns and Differentiation in Non-Metropolitan England*, p. 42, HMSO, London.

8 There could be all sorts of reasons, from differing social attitudes to the state of the road and rail networks, but the main factor is probably population density: commuters in much of highland Scotland and south western France would have to travel long distances to get to major population centres. On the other hand, the case of small German farmers going to work in factories is well known.

Chapter 9: The historical background to agricultural policy

1 The cost of getting it to the ports on the eastern seaboard of the USA decreased as a result of improvements in rail transport, and the cost of transporting it across the sea fell as more steamships became available.
2 There is a greater proportion of big farms in the UK.
3 People in Britain felt that they were buying food from countries which had largely been populated by the British. Many people would have known somebody who had moved to the colonies to farm, and so buying imports would be seen as a way of supporting them. But this may just have been the reaction of the chattering classes. It is also worth remembering that many imported products were cheaper than home-produced foodstuffs, and the less well-off would probably care as much about having enough to eat as about where it came from.
4 While not immune to criticism (especially about its cost to the taxpayer), it was a system which kept the UK market open to world trade, maintained an incentive to sell high quality products (if a farmer received more than the average market price his deficiency payment was not reduced), and left those who could best afford it – taxpayers, as opposed to food consumers, who include those with low or no incomes – to pay the cost of farm income support.

5   This is one of the great unanswered questions of modern agricultural economics. On the one hand, there would have been less of an incentive to produce more without price or income support, and so research stations and the ancillary industries have been encouraged to put resources into research, knowing that there would be a market for new varieties, fertilisers and fungicides. Farmers have been encouraged to spend more on inputs by the knowledge that they were receiving prices which were guaranteed in some way. On the other hand, scientists still had to come up with the technical changes and there seems to be no simple relationship between research expenditure and the rate of discovery, although of course agriculture is not the only industry which has seen rapid technical change in the second half of the twentieth century.

6   Because the UK was their main market.

7   The Country Landowners' Association (CLA) looks after the concerns of landowners, and the Transport and General Workers Union (TGWU) has a section for farm workers. Until the 1980s, farm workers had their own union, the National Union of Agricultural and Allied Workers, but as their numbers decreased they took the view that they would have more influence as members of one of the biggest trade unions.

8   The reasons listed in the text are that producer groups are more powerful than consumer and environmental groups, that the NFU gets the support of ancillary industry pressure groups, and that it has a close relationship with a government ministry which perceives its role as one of supporting the industry.

9   The reason suggested in the text is that public attitudes have changed now that a food shortage has been replaced by food surpluses, coupled with public perceptions of the environmental effects of modern agricultural methods.

10   COPA, to put it as crudely as possible, is likely to be effective because of its close relationship with the European Commission, but lacks effectiveness because it is a coalition of many farmers' organisations which will each have different objectives and so might find it difficult to agree on points of detail. It would be possible to produce a much more detailed discussion than this, but not in the space available here.

Chapter 10: The CAP and how it works

1 I; 2 E; 3 J; 4 H; 5 B; 6 D; 7 G; 8 A; 9 C; 10 F.

11

Enterprise	Price support by	Imports controlled by	Compensation payments
Cereals	Intervention buying for cereals of prescribed quality	Tariffs on imports	Arable Area Payments
Oilseeds	None	No import barriers	Arable Area Payments
Protein crops	None	No import barriers	Arable Area Payments
Linseed, hemp, flax	None	No import barriers	Arable Area Payments
Sugar	Intervention buying limited by quotas	Import levy	None
Potatoes	No CAP regime	No CAP regime	No CAP regime
Milk and milk products	Intervention prices for butter and skimmed milk powder	Import levy	None
Beef and veal	Intervention buying for limited quantities	Customs duties	Beef Special Premium Scheme Suckler Cow Premium Scheme
Sheepmeat	Basic price and aids to private storage	—	Sheep Annual Premium Scheme
Pigmeat	Basic price and aids to private storage	Sluicegate prices, basic and supplementary levies	None
Fruit and vegetables	Compensation for withdrawing produce from market paid to producer organisations	Common customs tariffs	None

12　If the UK becomes less self-sufficient in food products, imports will *rise* and the *supply curve* for pounds sterling on the international money markets will move to the *right*, so that the exchange rate of the pound will *fall*.

13   If interest rates in the UK rise, foreign investors will find it more desirable to put their money into the UK. The *demand curve* for pounds sterling on the international money markets will move to the *right* so that the exchange rate of the pound will *rise.*

14   If the rate of inflation in the UK rises, foreign investors will worry that the value of their holdings will decrease and so will find it less desirable to put their money into the UK. The *demand curve* for pounds sterling on the international money markets will move to the *left* so that the exchange rate of the pound will *fall.*

15   (a)   £98

    (b)   £93

16   Yes, the monetary gap is now:

$$\frac{(0.919585 - 0.94)}{0.919585} \times 100 = -2.22 \text{ (i.e. it is outside the boundary value of } -2\%)$$

so the process will have to be repeated to produce a new green rate of

$$0.94 \times \frac{100}{[100 - (2.22/2)]} = 0.92968$$

which reduces the monetary gap to –1.11. This is allowable unless there are positive monetary gaps which would produce a spread of more than 5 percentage points.

17 C; **18** D; **19** B; **20** D; **21** A.

INDEX